STUDENT SOLUTIONS MANUAL
FOR ZILL AND CULLEN'S

DIFFERENTIAL EQUATIONS
WITH BOUNDARY-VALUE
PROBLEMS

5TH EDITION

WARREN S. WRIGHT
Loyola Marymount University

CAROL D. WRIGHT

BROOKS/COLE

TM

THOMSON LEARNING

Australia • Canada • Mexico • Singapore • Spain • United Kingdom • United States

BROOKS/COLE

THOMSON LEARNING

Assistant Editor: *Carol Ann Benedict*
Editorial Assistant: *Daniel Thiem*
Marketing Manager: *Karin Sandberg*
Marketing Assistant: *Laurie Davidson*
Production Coordinator: *Dorothy Bell*

Cover Design: *Roy R. Neuhaus*
Cover Photo: *The Stock Market/*
Cameron Davidson
Print Buyer: *Micky Lawler*
Printing and Binding: *Webcom Limited*

For more information about this or any other Brooks/Cole product, contact:
BROOKS/COLE
511 Forest Lodge Road
Pacific Grove, CA 93950 USA
www.brookscole.com
1-800-423-0563 (Thomson Learning Academic Resource Center)

For permission to use material from this work, contact us by
Web: www.thomsonrights.com
fax: 1-800-730-2215
phone: 1-800-730-2214

Printed in Canada

10   9   8   7   6   5   4   3

ISBN 0-534-38003-4

Table of Contents

1 Introduction to Differential Equations

Exercises 1.1

3. The differential equation is first-order. Writing it in the form $x(dy/dx) + y^2 = 1$, we see that it is nonlinear in y because of y^2. However, writing it in the form $(y^2 - 1)(dx/dy) + x = 0$, we see that it is linear in x.

6. Second-order; nonlinear because of $\cos(r + u)$

9. Third-order; linear

12. From $y = \frac{6}{5} - \frac{6}{5}e^{-20t}$ we obtain $dy/dt = 24e^{-20t}$, so that

$$\frac{dy}{dt} + 20y = 24e^{-20t} + 20\left(\frac{6}{5} - \frac{6}{5}e^{-20t}\right) = 24.$$

15. Writing $\ln(2X - 1) - \ln(X - 1) = t$ and differentiating implicitly we obtain

$$\frac{2}{2X - 1}\frac{dX}{dt} - \frac{1}{X - 1}\frac{dX}{dt} = 1$$

$$\left(\frac{2}{2X - 1} - \frac{1}{X - 1}\right)\frac{dX}{dt} = 1$$

$$\frac{2X - 2 - 2X + 1}{(2X - 1)(X - 1)}\frac{dX}{dt} = 1$$

$$\frac{dX}{dt} = -(2X - 1)(X - 1) = (X - 1)(1 - 2X).$$

Exponentiating both sides of the implicit solution we obtain

$$\frac{2X - 1}{X - 1} = e^t \implies 2X - 1 = Xe^t - e^t \implies (e^t - 1) = (e^t - 2)X \implies X = \frac{e^t - 1}{e^t - 2}.$$

Solving $e^t - 2 = 0$ we get $t = \ln 2$. Thus, the solution is defined on $(-\infty, \ln 2)$ or on $(\ln 2, \infty)$. The graph of the solution defined on $(-\infty, \ln 2)$ is dashed, and the graph of the solution defined on $(\ln 2, \infty)$ is solid.

18. Differentiating $y = e^{-x^2}\int_0^x e^{t^2}\,dt + c_1 e^{-x^2}$ we obtain

$$y' = e^{-x^2}e^{x^2} - 2xe^{-x^2}\int_0^x e^{t^2}\,dt - 2c_1 xe^{-x^2} = 1 - 2xe^{-x^2}\int_0^x e^{t^2}\,dt - 2c_1 xe^{-x^2}.$$

Substituting into the differential equation, we have

$$y' + 2xy = 1 - 2xe^{-x^2}\int_0^x e^{t^2}\,dt - 2c_1 xe^{-x^2} + 2xe^{-x^2}\int_0^x e^{t^2}\,dt + 2c_1 xe^{-x^2} = 1.$$

21. (a) From $\phi_1 = x^2$ we obtain $\phi_1' = 2x$, so

$$x\phi_1' - 2\phi_1 = x(2x) - 2x^2 = 0.$$

From $\phi_2 = -x^2$ we obtain $\phi_2' = -2x$, so

$$x\phi_2' - 2\phi_2 = x(-2x) - 2(-x^2) = 0.$$

Thus, ϕ_1 and ϕ_2 are solutions of the differential equation on $(-\infty, \infty)$.

(b) From $y = \begin{cases} -x^2, & x < 0 \\ x^2, & x \geq 0 \end{cases}$ we obtain $y' = \begin{cases} -2x, & x < 0 \\ 2x, & x \geq 0 \end{cases}$ so that $xy' - 2y = 0$.

24. (a) An interval on which $\tan 5t$ is continuous is $-\pi/2 < 5t < \pi/2$, so $5\tan 5t$ will be a solution on $(-\pi/10, \pi/10)$.

(b) For $(1 - \sin t)^{-1/2}$ to be continuous we must have $1 - \sin t > 0$ or $\sin t < 1$. Thus, $(1 - \sin t)^{-1/2}$ will be a solution on $(\pi/2, 5\pi/2)$.

27. From $x = e^{-2t} + 3e^{6t}$ and $y = -e^{-2t} + 5e^{6t}$ we obtain

$$\frac{dx}{dt} = -2e^{-2t} + 18e^{6t} \quad \text{and} \quad \frac{dy}{dt} = 2e^{-2t} + 30e^{6t}.$$

Then

$$x + 3y = (e^{-2t} + 3e^{6t}) + 3(-e^{-2t} + 5e^{6t})$$

$$= -2e^{-2t} + 18e^{6t} = \frac{dx}{dt}$$

and

$$5x + 3y = 5(e^{-2t} + 3e^{6t}) + 3(-e^{-2t} + 5e^{6t})$$

$$= 2e^{-2t} + 30e^{6t} = \frac{dy}{dt}.$$

Exercises 1.2

3. Using $x' = -c_1 \sin t + c_2 \cos t$ we obtain $c_1 = -1$ and $c_2 = 8$. The solution is $x = -\cos t + 8\sin t$.

6. Using $x' = -c_1 \sin t + c_2 \cos t$ we obtain

$$\frac{\sqrt{2}}{2}c_1 + \frac{\sqrt{2}}{2}c_2 = \sqrt{2}$$

$$-\frac{\sqrt{2}}{2}c_1 + \frac{\sqrt{2}}{2}c_2 = 2\sqrt{2}.$$

Solving we find $c_1 = -1$ and $c_2 = 3$. The solution is $x = -\cos t + 3\sin t$.

9. From the initial conditions we obtain

$$c_1 e^{-1} + c_2 e = 5$$

$$c_1 e^{-1} - c_2 e = -5.$$

Solving we get $c_1 = 0$ and $c_2 = 5e^{-1}$. A solution of the initial-value problem is $y = 5e^{-x-1}$.

12. Two solutions are $y = 0$ and $y = x^2$. (Also, any constant multiple of x^2 is a solution.)

15. For $f(x,y) = \dfrac{y}{x}$ we have $\dfrac{\partial f}{\partial y} = \dfrac{1}{x}$. Thus the differential equation will have a unique solution in any region where $x \neq 0$.

18. For $f(x,y) = \dfrac{x^2}{1+y^3}$ we have $\dfrac{\partial f}{\partial y} = \dfrac{-3x^2 y^2}{(1+y^3)^2}$. Thus the differential equation will have a unique solution in any region where $y \neq -1$.

21. The differential equation has a unique solution at $(1,4)$.

24. The differential equation is not guaranteed to have a unique solution at $(-1,1)$.

27. **(a)** Since $\dfrac{d}{dt}\left(-\dfrac{1}{t+c}\right) = \dfrac{1}{(t+c)^2} = y^2$, we see that $y = -\dfrac{1}{t+c}$ is a solution of the differential equation.

(b) Solving $y(0) = -1/c = 1$ we obtain $c = -1$ and $y = 1/(1-t)$. Solving $y(0) = -1/c = -1$ we obtain $c = 1$ and $y = -1/(1+t)$. Being sure to include $t = 0$, we see that the interval of existence of $y = 1/(1-t)$ is $(-\infty, 1)$, while the interval of existence of $y = -1/(1+t)$ is $(-1, \infty)$.

(c) Solving $y(0) = -1/c = y_0$ we obtain $c = -1/y_0$ and

$$y = -\dfrac{1}{-1/y_0 + t} = \dfrac{y_0}{1 - y_0 t}, \quad y_0 \neq 0.$$

Since we must have $-1/y_0 + t \neq 0$, the largest interval of existence (which must contain 0) is either $(-\infty, 1/y_0)$ when $y_0 > 0$ or $(1/y_0, \infty)$ when $y_0 < 0$.

(d) By inspection we see that $y = 0$ is a solution on $(-\infty, \infty)$.

Exercises 1.3

3. Let b be the rate of births and d the rate of deaths. Then $b = k_1 P$ and $d = k_2 P^2$. Since $dP/dt = b-d$, the differential equation is $dP/dt = k_1 P - k_2 P^2$.

6. By inspecting the graph we take T_m to be $T_m(t) = 80 - 30 \cos \pi t/12$. Then the temperature of the body at time t is determined by the differential equation

$$\frac{dT}{dt} = k \left[T - \left(80 - 30 \cos \frac{\pi}{12} t \right) \right], \quad t > 0.$$

9. The rate at which salt is leaving the tank is

$$(3 \text{ gal/min}) \cdot \left(\frac{A}{300} \text{ lb/gal} \right) = \frac{A}{100} \text{ lb/min}.$$

Thus $dA/dt = A/100$.

12. The volume of water in the tank at time t is $V = \frac{1}{3}\pi r^2 h = \frac{1}{3} A_w h$. Using the formula from Problem 11 for the volume of water leaving the tank we see that the differential equation is

$$\frac{dh}{dt} = \frac{3}{A_w} \frac{dV}{dt} = \frac{3}{A_w} (-cA_h \sqrt{2gh}) = -\frac{3cA_h}{A_w} \sqrt{2gh}.$$

Using $A_h = \pi(2/12)^2 = \pi/36$, $g = 32$, and $c = 0.6$, this becomes

$$\frac{dh}{dt} = -\frac{3(0.6)\pi/36}{A_w} \sqrt{64h} = -\frac{0.4\pi}{A_w} h^{1/2}.$$

To find A_w we let r be the radius of the top of the water. Then $r/h = 8/20$, so $r = 2h/5$ and $A_w = \pi(2h/5)^2 = 4\pi h^2/25$. Thus

$$\frac{dh}{dt} = -\frac{0.4\pi}{4\pi h^2/25} h^{1/2} = -2.5 h^{-3/2}.$$

15. From Newton's second law we obtain $m \dfrac{dv}{dt} = -kv^2 + mg$.

18. From Problem 17, without a damping force, the differential equation is $m\, d^2 x/dt^2 = -kx$. With a damping force proportional to velocity the differential equation becomes

$$m \frac{d^2 x}{dt^2} = -kx - \beta \frac{dx}{dt} \quad \text{or} \quad m \frac{d^2 x}{dt^2} + \beta \frac{dx}{dt} + kx = 0.$$

21. From $g = k/R^2$ we find $k = gR^2$. Using $a = d^2 r/dt^2$ and the fact that the positive direction is upward we get

$$\frac{d^2 r}{dt^2} = -a = -\frac{k}{r^2} = -\frac{gR^2}{r^2} \quad \text{or} \quad \frac{d^2 r}{dt^2} + \frac{gR^2}{r^2} = 0.$$

24. The differential equation is $\dfrac{dA}{dt} = k_1(M - A) - k_2 A$.

4

27. We see from the figure that $2\theta + \alpha = \pi$. Thus

$$\frac{y}{-x} = \tan\alpha = \tan(\pi - 2\theta) = -\tan 2\theta = -\frac{2\tan\theta}{1 - \tan^2\theta}.$$

Since the slope of the tangent line is $y' = \tan\theta$ we have $y/x = 2y'[1 - (y')^2]$ or $y - y(y')^2 = 2xy'$, which is the quadratic equation $y(y')^2 + 2xy' - y = 0$ in y'. Using the quadratic formula we get

$$y' = \frac{-2x \pm \sqrt{4x^2 + 4y^2}}{2y} = \frac{-x \pm \sqrt{x^2 + y^2}}{y}.$$

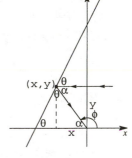

Since $dy/dx > 0$, the differential equation is

$$\frac{dy}{dx} = \frac{-x + \sqrt{x^2 + y^2}}{y} \qquad \text{or} \qquad y\frac{dy}{dx} - \sqrt{x^2 + y^2} + x = 0.$$

Chapter 1 Review Exercises

3. $\dfrac{d}{dx}(c_1\cos kx + c_2\sin kx) = -kc_1\sin kx + kc_2\cos kx;$

$\dfrac{d^2}{dx^2}(c_1\cos kx + c_2\sin kx) = -k^2c_1\cos kx - k^2c_2\sin kx = -k^2(c_1\cos kx + c_2\sin kx);$

$\dfrac{d^2y}{dx^2} = -k^2y \quad\text{or}\quad \dfrac{d^2y}{dx^2} + k^2y = 0$

6. $y' = -c_1e^x\sin x + c_1e^x\cos x + c_2e^x\cos x + c_2e^x\sin x;$

$y'' = -c_1e^x\cos x - c_1e^x\sin x - c_1e^x\sin x + c_1e^x\cos x - c_2e^x\sin x + c_2e^x\cos x + c_2e^x\cos x + c_2e^x\sin x$

$\quad = -2c_1e^x\sin x + 2c_2e^x\cos x;$

$y'' - 2y' = -2c_1e^x\cos x - 2c_2e^x\sin x = -2y; \qquad y'' - 2y' + 2y = 0$

9. b

12. a,b,d

15. The slope of the tangent line at (x, y) is y', so the differential equation is $y' = x^2 + y^2$.

18. (a) Differentiating $y^2 - 2y = x^2 - x + c$ we obtain $2yy' - 2y' = 2x - 1$ or $(2y - 2)y' = 2x - 1$.

(b) Setting $x = 0$ and $y = 1$ in the solution we have $1 - 2 = 0 - 0 + c$ or $c = -1$. Thus, a solution of the initial-value problem is $y^2 - 2y = x^2 - x - 1$.

(c) Solving $y^2 - 2y - (x^2 - x - 1) = 0$ by the quadratic formula we get $y = (2 \pm \sqrt{4 + 4(x^2 - x - 1)})/2$
$= 1 \pm \sqrt{x^2 - x} = 1 \pm \sqrt{x(x - 1)}$. Since $x(x - 1) \geq 0$ for $x \leq 0$ or $x \geq 1$, we see that neither

$y = 1 + \sqrt{x(x-1)}$ nor $y = 1 - \sqrt{x(x-1)}$ is differentiable at $x = 0$. Thus, both functions are solutions of the differential equation, but neither is a solution of the initial-value problem.

21. Differentiating $y = \sin(\ln x)$ we obtain $y' = \cos(\ln x)/x$ and $y'' = -[\sin(\ln x) + \cos(\ln x)]/x^2$. Then

$$x^2 y'' + xy' + y = x^2\left(-\frac{\sin(\ln x) + \cos(\ln x)}{x^2}\right) + x\frac{\cos(\ln x)}{x} + \sin(\ln x) = 0.$$

24. The differential equation is

$$\frac{dh}{dt} = -\frac{cA_0}{A_w}\sqrt{2gh}.$$

Using $A_0 = \pi(1/24)^2 = \pi/576$, $A_w = \pi(2)^2 = 4\pi$, and $g = 32$, this becomes

$$\frac{dh}{dt} = -\frac{c\pi/576}{4\pi}\sqrt{64h} = \frac{c}{288}\sqrt{h}.$$

2 First-Order Differential Equations

3. **6.**

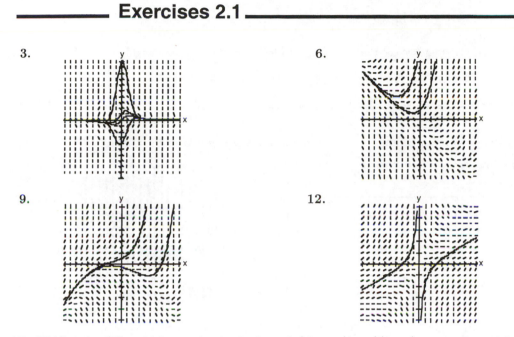

9. **12.**

15. Writing the differential equation in the form $dy/dx = y(1-y)(1+y)$ we see that critical points are located at $y = -1$, $y = 0$, and $y = 1$. The phase portrait is shown below.

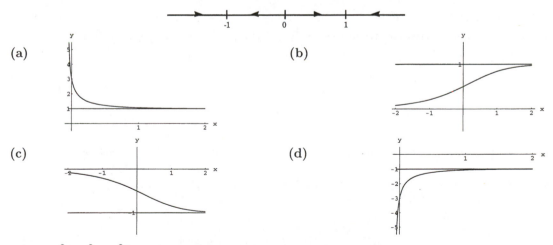

(a)

(b)

(c)

(d)

18. Solving $y^2 - y^3 = y^2(1-y) = 0$ we obtain the critical points 0 and 1.

From the phase portrait we see that 1 is asymptotically stable and 0 is semi-stable.

21. Solving $y^2(4 - y^2) = y^2(2 - y)(2 + y) = 0$ we obtain the critical points -2, 0, and 2.

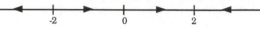

From the phase portrait we see that 2 is asymptotically stable, 0 is semi-stable, and -2 is unstable.

24. Solving $ye^y - 9y = y(e^y - 9) = 0$ we obtain the critical points 0 and $\ln 9$.

From the phase portrait we see that 0 is asymptotically stable and $\ln 9$ is unstable.

27. Critical points are $y = 0$ and $y = c$.

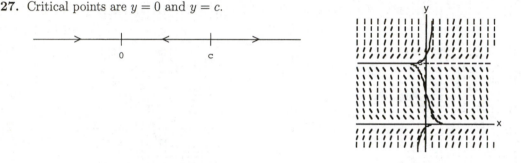

Exercises 2.2

In many of the following problems we will encounter an expression of the form $\ln|g(y)| = f(x) + c$. To solve for $g(y)$ we exponentiate both sides of the equation. This yields $|g(y)| = e^{f(x)+c} = e^c e^{f(x)}$ which implies $g(y) = \pm e^c e^{f(x)}$. Letting $c_1 = \pm e^c$ we obtain $g(y) = c_1 e^{f(x)}$.

3. From $dy = -e^{-3x}\, dx$ we obtain $y = \frac{1}{3}e^{-3x} + c$.

6. From $\dfrac{1}{y}\, dy = -2x\, dx$ we obtain $\ln|y| = -x^2 + c$ or $y = c_1 e^{-x^2}$.

9. From $\left(y + 2 + \dfrac{1}{y}\right) dy = x^2 \ln x\, dx$ we obtain $\dfrac{y^2}{2} + 2y + \ln|y| = \dfrac{x^3}{3}\ln|x| - \dfrac{1}{9}x^3 + c$.

12. From $2y\, dy = -\dfrac{\sin 3x}{\cos^3 3x}\, dx = -\tan 3x \sec^2 3x\, dx$ we obtain $y^2 = -\frac{1}{6}\sec^2 3x + c$.

15. From $\dfrac{1}{S}\, dS = k\, dr$ we obtain $S = ce^{kr}$.

18. From $\dfrac{1}{N}\, dN = \left(te^{t+2} - 1\right) dt$ we obtain $\ln|N| = te^{t+2} - e^{t+2} - t + c$.

21. From $x\,dx = \dfrac{1}{\sqrt{1-y^2}}\,dy$ we obtain $\frac{1}{2}x^2 = \sin^{-1} y + c$ or $y = \sin\left(\dfrac{x^2}{2} + c_1\right)$.

24. From $\dfrac{1}{y^2 - 1}\,dy = \dfrac{1}{x^2 - 1}\,dx$ or $\dfrac{1}{2}\left(\dfrac{1}{y-1} - \dfrac{1}{y+1}\right)dy = \dfrac{1}{2}\left(\dfrac{1}{x-1} - \dfrac{1}{x+1}\right)dx$ we obtain

$\ln|y-1| - \ln|y+1| = \ln|x-1| - \ln|x+1| + \ln c$ or $\dfrac{y-1}{y+1} = \dfrac{c(x-1)}{x+1}$. Using $y(2) = 2$ we find

$c = 1$. The solution of the initial-value problem is $\dfrac{y-1}{y+1} = \dfrac{x-1}{x+1}$ or $y = x$.

27. Separating variables and integrating we obtain

$$\frac{dx}{\sqrt{1-x^2}} - \frac{dy}{\sqrt{1-y^2}} = 0 \quad \text{and} \quad \sin^{-1} x - \sin^{-1} y = c.$$

Setting $x = 0$ and $y = \sqrt{3}/2$ we obtain $c = -\pi/3$. Thus, an implicit solution of the initial-value problem is $\sin^{-1} x - \sin^{-1} y = \pi/3$. Solving for y and using a trigonometric identity we get

$$y = \sin\left(\sin^{-1} x + \frac{\pi}{3}\right) = x\cos\frac{\pi}{3} + \sqrt{1-x^2}\,\sin\frac{\pi}{3} = \frac{x}{2} + \frac{\sqrt{3}\sqrt{1-x^2}}{2}.$$

30. From $\left(\dfrac{1}{y-1} + \dfrac{-1}{y}\right)dy = \dfrac{1}{x}\,dx$ we obtain $\ln|y-1| - \ln|y| = \ln|x| + c$ or $y = \dfrac{1}{1 - c_1 x}$.

Another solution is $y = 0$.

(a) If $y(0) = 1$ then $y = 1$.

(b) If $y(0) = 0$ then $y = 0$.

(c) If $y(1/2) = 1/2$ then $y = \dfrac{1}{1 + 2x}$.

(d) Setting $x = 2$ and $y = \frac{1}{4}$ we obtain

$$\frac{1}{4} = \frac{1}{1 - c_1(2)}, \quad 1 - 2c_1 = 4, \quad \text{and} \quad c_1 = -\frac{3}{2}.$$

Thus, $y = \dfrac{1}{1 + \frac{3}{2}x} = \dfrac{2}{2 + 3x}$.

33. The singular solution $y = 1$ satisfies the initial-value problem.

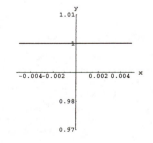

9

36. Separating variables we obtain $\dfrac{dy}{(y-1)^2 - 0.01} = dx$. Then

$$5\ln\left|\frac{10y-11}{10y-9}\right| = x + c.$$

Setting $x = 0$ and $y = 1$ we obtain $c = 5\ln 1 = 0$. The solution is

$$5\ln\left|\frac{10y-11}{10y-9}\right| = x.$$

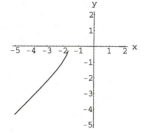

39. (a) Separating variables we have $2y\,dy = (2x+1)dx$. Integrating gives $y^2 = x^2 + x + c$. When $y(-2) = -1$ we find $c = -1$, so $y^2 = x^2 + x - 1$ and $y = -\sqrt{x^2 + x - 1}$. The negative square root is chosen because of the initial condition.

(b) The interval of definition appears to be approximately $(-\infty, -1.65)$.

(c) Solving $x^2 + x - 1 = 0$ we get $x = -\frac{1}{2} \pm \frac{1}{2}\sqrt{5}$, so the exact interval of definition is $(-\infty, -\frac{1}{2} - \frac{1}{2}\sqrt{5})$.

Exercises 2.3

3. For $y' + y = e^{3x}$ an integrating factor is $e^{\int dx} = e^x$ so that $\dfrac{d}{dx}[e^x y] = e^{4x}$ and $y = \frac{1}{4}e^{3x} + ce^{-x}$ for $-\infty < x < \infty$. The transient term is ce^{-x}.

6. For $y' + 2xy = x^3$ an integrating factor is $e^{\int 2x\,dx} = e^{x^2}$ so that $\dfrac{d}{dx}\left[e^{x^2} y\right] = x^3 e^{x^2}$ and $y = \frac{1}{2}x^2 - \frac{1}{2} + ce^{-x^2}$ for $-\infty < x < \infty$. The transient term is ce^{-x^2}.

9. For $y' - \dfrac{1}{x}y = x\sin x$ an integrating factor is $e^{-\int(1/x)dx} = \dfrac{1}{x}$ so that $\dfrac{d}{dx}\left[\dfrac{1}{x}y\right] = \sin x$ and $y = cx - x\cos x$ for $0 < x < \infty$.

12. For $y' - \dfrac{x}{(1+x)}y = x$ an integrating factor is $e^{-\int[x/(1+x)]dx} = (x+1)e^{-x}$ so that $\dfrac{d}{dx}\left[(x+1)e^{-x}y\right] = x(x+1)e^{-x}$ and $y = -x - \dfrac{2x+3}{x+1} + \dfrac{ce^x}{x+1}$ for $-1 < x < \infty$.

15. For $\dfrac{dx}{dy} - \dfrac{4}{y}x = 4y^5$ an integrating factor is $e^{-\int (4/y)\,dy} = y^{-4}$ so that $\dfrac{d}{dy}\left[y^{-4}x\right] = 4y$ and

$x = 2y^6 + cy^4$ for $0 < y < \infty$.

18. For $y' + (\cot x)y = \sec^2 x \csc x$ an integrating factor is $e^{\int \cot x\,dx} = \sin x$ so that $\dfrac{d}{dx}\left[(\sin x)\,y\right] = \sec^2 x$

and $y = \sec x + c\csc x$ for $0 < x < \pi/2$.

21. For $\dfrac{dr}{d\theta} + r\sec\theta = \cos\theta$ an integrating factor is $e^{\int \sec\theta\,d\theta} = \sec\theta + \tan\theta$ so that $\dfrac{d}{d\theta}\left[r(\sec\theta + \tan\theta)\right] =$

$1 + \sin\theta$ and $r(\sec\theta + \tan\theta) = \theta - \cos\theta + c$ for $-\pi/2 < \theta < \pi/2$.

24. For $y' + \dfrac{2}{x^2-1}y = \dfrac{x+1}{x-1}$ an integrating factor is $e^{\int [2/(x^2-1)]\,dx} = \dfrac{x-1}{x+1}$ so that $\dfrac{d}{dx}\left[\dfrac{x-1}{x+1}y\right] = 1$

and $(x-1)y = x(x+1) + c(x+1)$ for $-1 < x < 1$.

27. For $\dfrac{di}{dt} + \dfrac{R}{L}i = \dfrac{E}{L}$ an integrating factor is $e^{\int (R/L)\,dt} = e^{Rt/L}$ so that $\dfrac{d}{dt}\left[ie^{Rt/L}\right] = \dfrac{E}{L}e^{Rt/L}$ and

$i = \dfrac{E}{R} + ce^{-Rt/L}$ for $-\infty < t < \infty$. If $i(0) = i_0$ then $c = i_0 - E/R$ and $i = \dfrac{E}{R} + \left(i_0 - \dfrac{E}{R}\right)e^{-Rt/L}$.

30. For $y' + (\tan x)y = \cos^2 x$ an integrating factor is $e^{\int \tan x\,dx} = \sec x$ so that $\dfrac{d}{dx}\left[(\sec x)\,y\right] = \cos x$ and

$y = \sin x \cos x + c \cos x$ for $-\pi/2 < x < \pi/2$. If $y(0) = -1$ then $c = -1$ and $y = \sin x \cos x - \cos x$.

33. For $y' + 2xy = f(x)$ an integrating factor is e^{x^2} so that

$$ye^{x^2} = \begin{cases} \frac{1}{2}e^{x^2} + c_1, & 0 \le x \le 1; \\ c_2, & x > 1. \end{cases}$$

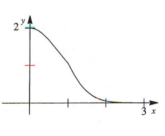

If $y(0) = 2$ then $c_1 = 3/2$ and for continuity we must have $c_2 = \frac{1}{2}e + \frac{3}{2}$ so that

$$y = \begin{cases} \frac{1}{2} + \frac{3}{2}e^{-x^2}, & 0 \le x \le 1; \\ \left(\frac{1}{2}e + \frac{3}{2}\right)e^{-x^2}, & x > 1. \end{cases}$$

36. An integrating factor for $y' - 2xy = 1$ is e^{-x^2}. Thus

$$\frac{d}{dx}\left[e^{-x^2}y\right] = e^{-x^2}$$

$$e^{-x^2}y = \int_0^x e^{-t^2}\,dt = \operatorname{erf}(x) + c$$

and

$$y = e^{x^2}\operatorname{erf}(x) + ce^{x^2}.$$

From $y(1) = 1$ we get $1 = e\operatorname{erf}(1) + ce$, so that $c = e^{-1} - \operatorname{erf}(1)$. Thus

$$y = e^{x^2}\operatorname{erf}(x) + (e^{-1} - \operatorname{erf}(1))e^{x^2} = e^{x^2-1} + e^{x^2}(\operatorname{erf}(x) - \operatorname{erf}(1)).$$

Exercises 2.4

3. Let $M = 5x + 4y$ and $N = 4x - 8y^3$ so that $M_y = 4 = N_x$. From $f_x = 5x + 4y$ we obtain $f = \frac{5}{2}x^2 + 4xy + h(y)$, $h'(y) = -8y^3$, and $h(y) = -2y^4$. The solution is $\frac{5}{2}x^2 + 4xy - 2y^4 = c$.

6. Let $M = 4x^3 - 3y\sin 3x - y/x^2$ and $N = 2y - 1/x + \cos 3x$ so that $M_y = -3\sin 3x - 1/x^2$ and $N_x = 1/x^2 - 3\sin 3x$. The equation is not exact.

9. Let $M = y^3 - y^2\sin x - x$ and $N = 3xy^2 + 2y\cos x$ so that $M_y = 3y^2 - 2y\sin x = N_x$. From $f_x = y^3 - y^2\sin x - x$ we obtain $f = xy^3 + y^2\cos x - \frac{1}{2}x^2 + h(y)$, $h'(y) = 0$, and $h(y) = 0$. The solution is $xy^3 + y^2\cos x - \frac{1}{2}x^2 = c$.

12. Let $M = 3x^2y + e^y$ and $N = x^3 + xe^y - 2y$ so that $M_y = 3x^2 + e^y = N_x$. From $f_x = 3x^2y + e^y$ we obtain $f = x^3y + xe^y + h(y)$, $h'(y) = -2y$, and $h(y) = -y^2$. The solution is $x^3y + xe^y - y^2 = c$.

15. Let $M = x^2y^3 - 1/\left(1 + 9x^2\right)$ and $N = x^3y^2$ so that $M_y = 3x^2y^2 = N_x$. From $f_x = x^2y^3 - 1/\left(1 + 9x^2\right)$ we obtain $f = \frac{1}{3}x^3y^3 - \frac{1}{3}\arctan(3x) + h(y)$, $h'(y) = 0$, and $h(y) = 0$. The solution is $x^3y^3 - \arctan(3x) = c$.

18. Let $M = 2y\sin x\cos x - y + 2y^2e^{xy^2}$ and $N = -x + \sin^2 x + 4xye^{xy^2}$ so that

$$M_y = 2\sin x\cos x - 1 + 4xy^3e^{xy^2} + 4ye^{xy^2} = N_x.$$

From $f_x = 2y\sin x\cos x - y + 2y^2e^{xy^2}$ we obtain $f = y\sin^2 x - xy + 2e^{xy^2} + h(y)$, $h'(y) = 0$, and $h(y) = 0$. The solution is $y\sin^2 x - xy + 2e^{xy^2} = c$.

21. Let $M = x^2 + 2xy + y^2$ and $N = 2xy + x^2 - 1$ so that $M_y = 2(x+y) = N_x$. From $f_x = x^2 + 2xy + y^2$ we obtain $f = \frac{1}{3}x^3 + x^2y + xy^2 + h(y)$, $h'(y) = -1$, and $h(y) = -y$. The general solution is $\frac{1}{3}x^3 + x^2y + xy^2 - y = c$. If $y(1) = 1$ then $c = \frac{4}{3}$ and the solution of the initial-value problem is $\frac{1}{3}x^3 + x^2y + xy^2 - y = \frac{4}{3}$.

24. Let $M = t/2y^4$ and $N = \left(3y^2 - t^2\right)/y^5$ so that $M_y = -2t/y^5 = N_t$. From $f_t = t/2y^4$ we obtain $f = \dfrac{t^2}{4y^4} + h(y)$, $h'(y) = \dfrac{3}{y^3}$, and $h(y) = -\dfrac{3}{2y^2}$. The general solution is $\dfrac{t^2}{4y^4} - \dfrac{3}{2y^2} = c$. If $y(1) = 1$ then $c = -5/4$ and the solution of the initial-value problem is $\dfrac{t^2}{4y^4} - \dfrac{3}{2y^2} = -\dfrac{5}{4}$.

27. Equating $M_y = 3y^2 + 4kxy^3$ and $N_x = 3y^2 + 40xy^3$ we obtain $k = 10$.

30. Let $M = \left(x^2 + 2xy - y^2\right)/\left(x^2 + 2xy + y^2\right)$ and $N = \left(y^2 + 2xy - x^2\right)/\left(y^2 + 2xy + x^2\right)$ so that $M_y = -4xy/(x+y)^3 = N_x$. From $f_x = \left(x^2 + 2xy + y^2 - 2y^2\right)/(x+y)^2$ we obtain $f = x + \dfrac{2y^2}{x+y} + h(y)$, $h'(y) = -1$, and $h(y) = -y$. The solution of the differential equation is $x^2 + y^2 = c(x+y)$.

33. We note that $(N_x - M_y)/M = 2/y$, so an integrating factor is $e^{\int 2dy/y} = y^2$. Let $M = 6xy^3$ and $N = 4y^3 + 9x^2y^2$ so that $M_y = 18xy^2 = N_x$. From $f_x = 6xy^3$ we obtain $f = 3x^2y^3 + h(y)$,

$h'(y) = 4y^3$, and $h(y) = y^4$. The solution of the differential equation is $3x^2y^3 + y^4 = c$.

36. We note that $(N_x - M_y)/M = -3/y$, so an integrating factor is $e^{-3\int dy/y} = 1/y^3$. Let

$$M = (y^2 + xy^3)/y^3 = 1/y + x$$

and

$$N = (5y^2 - xy + y^3 \sin y)/y^3 = 5/y - x/y^2 + \sin y,$$

so that $M_y = -1/y^2 = N_x$. From $f_x = 1/y + x$ we obtain $f = x/y + \frac{1}{2}x^2 + h(y)$, $h'(y) = 5/y + \sin y$, and $h(y) = 5 \ln|y| - \cos y$. The solution of the differential equation is $x/y + \frac{1}{2}x^2 + 5 \ln|y| - \cos y = c$.

39. (a) Implicitly differentiating $x^3 + 2x^2y + y^2 = c$ and solving for dy/dx we obtain

$$3x^2 + 2x^2 \frac{dy}{dx} + 4xy + 2y \frac{dy}{dx} = 0 \quad \text{and} \quad \frac{dy}{dx} = -\frac{3x^2 + 4xy}{2x^2 + 2y}.$$

Separating variables we get $(4xy + 3x^2)dx + (2y + 2x^2)dy = 0$.

(b) Setting $x = 0$ and $y = -2$ in $x^3 + 2x^2y + y^2 = c$ we find $c = 4$, and setting $x = y = 1$ we also find $c = 4$. Thus, both initial conditions determine the same implicit solution.

(c) Solving $x^3 + 2x^2y + y^2 = 4$ for y we get

$$y_1(x) = -x^2 - \sqrt{4 - x^3 + x^4}$$

and

$$y_2(x) = -x^2 + \sqrt{4 - x^3 + x^4}.$$

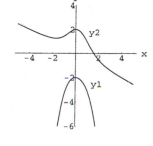

Exercises 2.5

3. Letting $x = vy$ we have

$$vy(v\,dy + y\,dv) + (y - 2vy)\,dy = 0$$

$$vy\,dv + \left(v^2 - 2v + 1\right)dy = 0$$

$$\frac{v\,dv}{(v-1)^2} + \frac{dy}{y} = 0$$

$$\ln|v-1| - \frac{1}{v-1} + \ln|y| = c$$

$$\ln\left|\frac{x}{y} - 1\right| - \frac{1}{x/y - 1} + \ln y = c$$

$$(x - y)\ln|x - y| - y = c(x - y).$$

Exercises 2.5

6. Letting $y = ux$ we have

$$\left(u^2x^2 + ux^2\right)dx + x^2(u\,dx + x\,du) = 0$$

$$\left(u^2 + 2u\right)dx + x\,du = 0$$

$$\frac{dx}{x} + \frac{du}{u(u+2)} = 0$$

$$\ln|x| + \frac{1}{2}\ln|u| - \frac{1}{2}\ln|u+2| = c$$

$$\frac{x^2u}{u+2} = c_1$$

$$x^2\frac{y}{x} = c_1\left(\frac{y}{x} + 2\right)$$

$$x^2y = c_1(y + 2x).$$

9. Letting $y = ux$ we have

$$-ux\,dx + (x + \sqrt{u}\,x)(u\,dx + x\,du) = 0$$

$$(x + x\sqrt{u})\,du + u^{3/2}\,dx = 0$$

$$\left(u^{-3/2} + \frac{1}{u}\right)du + \frac{dx}{x} = 0$$

$$-2u^{-1/2} + \ln|u| + \ln|x| = c$$

$$\ln|y/x| + \ln|x| = 2\sqrt{x/y} + c$$

$$y(\ln|y| - c)^2 = 4x.$$

12. Letting $y = ux$ we have

$$\left(x^2 + 2u^2x^2\right)dx - ux^2(u\,dx + x\,du) = 0$$

$$\left(1 + u^2\right)dx - ux\,du = 0$$

$$\frac{dx}{x} - \frac{u\,du}{1+u^2} = 0$$

$$\ln|x| - \frac{1}{2}\ln\left(1 + u^2\right) = c$$

$$\frac{x^2}{1+u^2} = c_1$$

$$x^4 = c_1\left(y^2 + x^2\right).$$

14

Using $y(-1) = 1$ we find $c_1 = 1/2$. The solution of the initial-value problem is $2x^4 = y^2 + x^2$.

15. From $y' + \dfrac{1}{x}y = \dfrac{1}{x}y^{-2}$ and $w = y^3$ we obtain $\dfrac{dw}{dx} + \dfrac{3}{x}w = \dfrac{3}{x}$. An integrating factor is x^3 so that $x^3 w = x^3 + c$ or $y^3 = 1 + cx^{-3}$.

18. From $y' - \left(1 + \dfrac{1}{x}\right)y = y^2$ and $w = y^{-1}$ we obtain $\dfrac{dw}{dx} + \left(1 + \dfrac{1}{x}\right)w = -1$. An integrating factor is xe^x so that $xe^x w = -xe^x + e^x + c$ or $y^{-1} = -1 + \dfrac{1}{x} + \dfrac{c}{x}e^{-x}$.

21. From $y' - \dfrac{2}{x}y = \dfrac{3}{x^2}y^4$ and $w = y^{-3}$ we obtain $\dfrac{dw}{dx} + \dfrac{6}{x}w = -\dfrac{9}{x^2}$. An integrating factor is x^6 so that $x^6 w = -\dfrac{9}{5}x^5 + c$ or $y^{-3} = -\dfrac{9}{5}x^{-1} + cx^{-6}$. If $y(1) = \dfrac{1}{2}$ then $c = \dfrac{49}{5}$ and $y^{-3} = -\dfrac{9}{5}x^{-1} + \dfrac{49}{5}x^{-6}$.

24. Let $u = x + y$ so that $du/dx = 1 + dy/dx$. Then $\dfrac{du}{dx} - 1 = \dfrac{1-u}{u}$ or $u\,du = dx$. Thus $\dfrac{1}{2}u^2 = x + c$ or $u^2 = 2x + c_1$, and $(x + y)^2 = 2x + c_1$.

27. Let $u = y - 2x + 3$ so that $du/dx = dy/dx - 2$. Then $\dfrac{du}{dx} + 2 = 2 + \sqrt{u}$ or $\dfrac{1}{\sqrt{u}}\,du = dx$. Thus $2\sqrt{u} = x + c$ and $2\sqrt{y - 2x + 3} = x + c$.

30. Let $u = 3x + 2y$ so that $du/dx = 3 + 2\,dy/dx$. Then $\dfrac{du}{dx} = 3 + \dfrac{2u}{u+2} = \dfrac{5u+6}{u+2}$ and $\dfrac{u+2}{5u+6}\,du = dx$. Now

$$\frac{u+2}{5u+6} = \frac{1}{5} + \frac{4}{25u+30}$$

so we have

$$\int \left(\frac{1}{5} + \frac{4}{25u+30}\right) du = dx$$

and $\dfrac{1}{5}u + \dfrac{4}{25}\ln|25u + 30| = x + c$. Thus

$$\frac{1}{5}(3x + 2y) + \frac{4}{25}\ln|75x + 50y + 30| = x + c.$$

Setting $x = -1$ and $y = -1$ we obtain $c = \dfrac{4}{5}\ln 95$. The solution is

$$\frac{1}{5}(3x + 2y) + \frac{4}{25}\ln|75x + 50y + 30| = x + \frac{4}{5}\ln 95$$

or

$$5y - 5x + 2\ln|75x + 50y + 30| = 10\ln 95.$$

15

Exercises 2.6

3. Separating variables and integrating, we have

$$\frac{dy}{y} = dx \quad \text{and} \quad \ln|y| = x + c.$$

Thus $y = c_1 e^x$ and, using $y(0) = 1$, we find $c = 1$, so $y = e^x$ is the solution of the initial-value problem.

h=0.1

x_n	y_n	True Value	Abs. Error	% Rel. Error
0.00	1.0000	1.0000	0.0000	0.00
0.10	1.1000	1.1052	0.0052	0.47
0.20	1.2100	1.2214	0.0114	0.93
0.30	1.3310	1.3499	0.0189	1.40
0.40	1.4641	1.4918	0.0277	1.86
0.50	1.6105	1.6487	0.0382	2.32
0.60	1.7716	1.8221	0.0506	2.77
0.70	1.9487	2.0138	0.0650	3.23
0.80	2.1436	2.2255	0.0820	3.68
0.90	2.3579	2.4596	0.1017	4.13
1.00	2.5937	2.7183	0.1245	4.58

h=0.05

x_n	y_n	True Value	Abs. Error	% Rel. Error
0.00	1.0000	1.0000	0.0000	0.00
0.05	1.0500	1.0513	0.0013	0.12
0.10	1.1025	1.1052	0.0027	0.24
0.15	1.1576	1.1618	0.0042	0.36
0.20	1.2155	1.2214	0.0059	0.48
0.25	1.2763	1.2840	0.0077	0.60
0.30	1.3401	1.3499	0.0098	0.72
0.35	1.4071	1.4191	0.0120	0.84
0.40	1.4775	1.4918	0.0144	0.96
0.45	1.5513	1.5683	0.0170	1.08
0.50	1.6289	1.6487	0.0198	1.20
0.55	1.7103	1.7333	0.0229	1.32
0.60	1.7959	1.8221	0.0263	1.44
0.65	1.8856	1.9155	0.0299	1.56
0.70	1.9799	2.0138	0.0338	1.68
0.75	2.0789	2.1170	0.0381	1.80
0.80	2.1829	2.2255	0.0427	1.92
0.85	2.2920	2.3396	0.0476	2.04
0.90	2.4066	2.4596	0.0530	2.15
0.95	2.5270	2.5857	0.0588	2.27
1.00	2.6533	2.7183	0.0650	2.39

6.

h=0.1

x_n	y_n
0.00	1.0000
0.10	1.1000
0.20	1.2220
0.30	1.3753
0.40	1.5735
0.50	1.8371

h=0.05

x_n	y_n
0.00	1.0000
0.05	1.0500
0.10	1.1053
0.15	1.1668
0.20	1.2360
0.25	1.3144
0.30	1.4039
0.35	1.5070
0.40	1.6267
0.45	1.7670
0.50	1.9332

9.

h=0.1

x_n	y_n
1.00	1.0000
1.10	1.0000
1.20	1.0191
1.30	1.0588
1.40	1.1231
1.50	1.2194

h=0.05

x_n	y_n
1.00	1.0000
1.05	1.0000
1.10	1.0049
1.15	1.0147
1.20	1.0298
1.25	1.0506
1.30	1.0775
1.35	1.1115
1.40	1.1538
1.45	1.2057
1.50	1.2696

12.

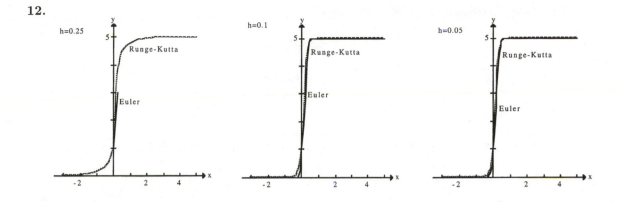

————— **Chapter 2 Review Exercises** —————

3. $\dfrac{dy}{dx} = (y-1)^2(y-3)^2$

6. The zero of f occurs at approximately 1.3. Since $P'(t) = f(P) > 0$ for $P < 1.3$ and $P'(t) = f(P) > 0$ for $P > 1.3$, $\lim_{t\to\infty} P(t) = 1.3$.

9. Separating variables we obtain

$$\cos^2 x\, dx = \frac{y}{y^2+1}\, dy \implies \frac{1}{2}x + \frac{1}{4}\sin 2x = \frac{1}{2}\ln\left(y^2+1\right) + c \implies 2x + \sin 2x = 2\ln\left(y^2+1\right) + c.$$

12. Write the differential equation in the form $(3y^2 + 2x)dx + (4y^2 + 6xy)dy = 0$. Letting $M = 3y^2 + 2x$ and $N = 4y^2 + 6xy$ we see that $M_y = 6y = N_x$ so the differential equation is exact. From $f_x = 3y^2 + 2x$ we obtain $f = 3xy^2 + x^2 + h(y)$. Then $f_y = 6xy + h'(y) = 4y^2 + 6xy$ and $h'(y) = 4y^2$ so $h(y) = \frac{4}{3}y^3$. The general solution is

$$3xy^2 + x^2 + \frac{4}{3}y^3 = c.$$

15. Write the equation in the form $\dfrac{dy}{dx} + \dfrac{8x}{x^2+4}y = \dfrac{2x}{x^2+4}$. An integrating factor is $\left(x^2+4\right)^4$, so

$$\frac{d}{dx}\left[\left(x^2+4\right)^4 y\right] = 2x\left(x^2+4\right)^3 \implies \left(x^2+4\right)^4 y = \frac{1}{4}\left(x^2+4\right)^4 + c \implies y = \frac{1}{4} + c\left(x^2+4\right)^{-4}.$$

17

18. Separating variables and integrating we have

$$\frac{dy}{y^2} = -2(t+1)\,dt$$

$$-\frac{1}{y} = -(t+1)^2 + c$$

$$y = \frac{1}{(t+1)^2 + c}.$$

The initial condition implies $c = -9$, so the solution of the initial-value problem is

$$y = \frac{1}{t^2 + 2t - 8} \quad \text{where} \quad -4 < t < 2.$$

21. The graph of $y_1(x)$ is the portion of the closed black curve lying in the fourth quadrant. Its interval of definition is approximately $(0.7, 4.3)$. The graph of $y_2(x)$ is the portion of the left-hand black curve lying in the third quadrant. Its interval of definition is $(-\infty, 0)$.

3 Modeling with First-Order Differential Equations

Exercises 3.1

3. Let $P = P(t)$ be the population at time t. From $dP/dt = kt$ and $P(0) = P_0 = 500$ we obtain $P = 500e^{kt}$. Using $P(10) = 575$ we find $k = \frac{1}{10}\ln 1.15$. Then $P(30) = 500e^{3\ln 1.15} \approx 760$ years.

6. From $dS/dt = rS$ we obtain $S = S_0 e^{rt}$ where $S(0) = S_0$.
    (a) If $S_0 = \$5000$ and $r = 5.75\%$ then $S(5) = \$6665.45$.
    (b) If $S(t) = \$10,000$ then $t = 12$ years.
    (c) $S \approx \$6651.82$

9. Setting $N(t) = 50$ in Problem 8 we obtain

$$50 = 100e^{kt} \implies kt = \ln\frac{1}{2} \implies t = \frac{\ln 1/2}{(1/6)\ln 0.97} \approx 136.5 \text{ hours.}$$

12. From Example 3, the amount of carbon present at time t is $A(t) = A_0 e^{-0.00012378t}$. Letting $t = 660$ and solving for A_0 we have $A(660) = A_0 e^{-0.0001237(660)} = 0.921553A_0$. Thus, approximately 92% of the original amount of C-14 remained in the cloth as of 1988.

15. Assume that $dT/dt = k(T - 100)$ so that $T = 100 + ce^{kt}$. If $T(0) = 20°$ and $T(1) = 22°$ then $c = -80$ and $k = \ln(39/40)$ so that $T(t) = 90°$ implies $t = 82.1$ seconds. If $T(t) = 98°$ then $t = 145.7$ seconds.

18. From $dA/dt = 0 - A/50$ we obtain $A = ce^{-t/50}$. If $A(0) = 30$ then $c = 30$ and $A = 30e^{-t/50}$.

21. From $\dfrac{dA}{dt} = 3 - \dfrac{4A}{100 + (6-4)t} = 3 - \dfrac{2A}{50 + t}$ we obtain $A = 50 + t + c(50 + t)^{-2}$. If $A(0) = 10$ then $c = -100{,}000$ and $A(30) = 64.38$ pounds.

24. Assume $L\,di/dt + Ri = E(t)$, $E(t) = E_0 \sin \omega t$, and $i(0) = i_0$ so that

$$i = \frac{E_0 R}{L^2\omega^2 + R^2}\sin \omega t - \frac{E_0 L\omega}{L^2\omega^2 + R^2}\cos \omega t + ce^{-Rt/L}.$$

Since $i(0) = i_0$ we obtain $c = i_0 + \dfrac{E_0 L\omega}{L^2\omega^2 + R^2}$.

27. For $0 \leq t \leq 20$ the differential equation is $20\,di/dt + 2i = 120$. An integrating factor is $e^{t/10}$, so $\dfrac{d}{dt}\left[e^{t/10}i\right] = 6e^{t/10}$ and $i = 60 + c_1 e^{-t/10}$. If $i(0) = 0$ then $c_1 = -60$ and $i = 60 - 60e^{-t/10}$.

For $t > 20$ the differential equation is $20\, di/dt + 2i = 0$ and $i = c_2 e^{-t/10}$.

At $t = 20$ we want $c_2 e^{-2} = 60 - 60 e^{-2}$ so that $c_2 = 60 \left(e^2 - 1\right)$. Thus

$$i(t) = \begin{cases} 60 - 60 e^{-t/10}, & 0 \leq t \leq 20; \\ 60 \left(e^2 - 1\right) e^{-t/10}, & t > 20. \end{cases}$$

30. (a) Integrating $d^2 s/dt^2 = -g$ we get $v(t) = ds/dt = -gt + c$. From $v(0) = 300$ we find $c = 300$, so the velocity is $v(t) = -32t + 300$.

(b) Integrating again and using $s(0) = 0$ we get $s(t) = -16t^2 + 300t$. The maximum height is attained when $v = 0$, that is, at $t_a = 9.375$. The maximum height will be $s(9.375) = 1406.25\,\text{ft}$.

33. (a) The differential equation is first-order, linear. Letting $b = k/\rho$, the integrating factor is $e^{\int 3b\, dt/(bt + r_0)} = (r_0 + bt)^3$. Then

$$\frac{d}{dt}[(r_0 + bt)^3 v] = g(r_0 + bt)^3 \quad \text{and} \quad (r_0 + bt)^3 v = \frac{g}{4b}(r_0 + bt)^4 + c.$$

The solution of the differential equation is $v(t) = (g/4b)(r_0 + bt) + c(r_0 + bt)^{-3}$. Using $v(0) = 0$ we find $c = -gr_0^4/4b$, so that

$$v(t) = \frac{g}{4b}(r_0 + bt) - \frac{gr_0^4}{4b(r_0 + bt)^3} = \frac{g\rho}{4k}\left(r_0 + \frac{k}{\rho}t\right) - \frac{g\rho r_0^4}{4k(r_0 + kt/\rho)^3}.$$

(b) Integrating $dr/dt = k/\rho$ we get $r = kt/\rho + c$. Using $r(0) = r_0$ we have $c = r_0$, so $r(t) = kt/\rho + r_0$.

(c) If $r = 0.007\,\text{ft}$ when $t = 10\,\text{s}$, then solving $r(10) = 0.007$ for k/ρ, we obtain $k/\rho = -0.0003$ and $r(t) = 0.01 - 0.0003t$. Solving $r(t) = 0$ we get $t = 33.3$, so the raindrop will have evaporated completely at 33.3 seconds.

36. The first equation can be solved by separation of variables. We obtain $x = c_1 e^{-\lambda_1 t}$. From $x(0) = x_0$ we obtain $c_1 = x_0$ and so $x = x_0 e^{-\lambda_1 t}$. The second equation then becomes

$$\frac{dy}{dt} = x_0 \lambda_1 e^{-\lambda_1 t} - \lambda_2 y \quad \text{or} \quad \frac{dy}{dt} + \lambda_2 y = x_0 \lambda_1 e^{-\lambda_1 t}$$

which is linear. An integrating factor is $e^{\lambda_2 t}$. Thus

$$\frac{d}{dt}\left[e^{\lambda_2 t} y\right] = x_0 \lambda_1 e^{-\lambda_1 t} e^{\lambda_2 t} = x_0 \lambda_1 e^{(\lambda_2 - \lambda_1)t}$$

$$e^{\lambda_2 t} y = \frac{x_0 \lambda_1}{\lambda_2 - \lambda_1} e^{(\lambda_2 - \lambda_1)t} + c_2$$

$$y = \frac{x_0 \lambda_1}{\lambda_2 - \lambda_1} e^{-\lambda_1 t} + c_2 e^{-\lambda_2 t}.$$

From $y(0) = y_0$ we obtain $c_2 = (y_0 \lambda_2 - y_0 \lambda_1 - x_0 \lambda_1)/(\lambda_2 - \lambda_1)$. The solution is

$$y = \frac{x_0 \lambda_1}{\lambda_2 - \lambda_1} e^{-\lambda_1 t} + \frac{y_0 \lambda_2 - y_0 \lambda_1 - x_0 \lambda_1}{\lambda_2 - \lambda_1} e^{-\lambda_2 t}.$$

Exercises 3.2

3. From $\dfrac{dP}{dt} = P\left(10^{-1} - 10^{-7}P\right)$ and $P(0) = 5000$ we obtain $P = \dfrac{500}{0.0005 + 0.0995e^{-0.1t}}$ so that $P \to 1{,}000{,}000$ as $t \to \infty$. If $P(t) = 500{,}000$ then $t = 52.9$ months.

6. From $\dfrac{dP}{dt} = P(a - b\ln P)$ we obtain $\dfrac{-1}{b}\ln|a - b\ln P| = t + c_1$ so that $P = e^{a/b}e^{-ce^{-bt}}$. If $P(0) = P_0$ then $c = \dfrac{a}{b} - \ln P_0$.

9. (a) The initial-value problem is $dh/dt = -8A_h\sqrt{h}/A_w$, $h(0) = H$. Separating variables and integrating we have

$$\frac{dh}{\sqrt{h}} = -\frac{8A_h}{A_w}\,dt \quad \text{and} \quad 2\sqrt{h} = -\frac{8A_h}{A_w}t + c.$$

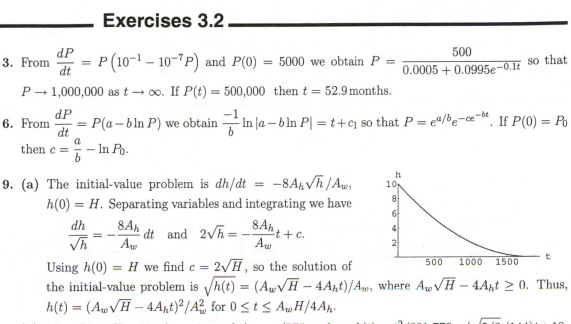

Using $h(0) = H$ we find $c = 2\sqrt{H}$, so the solution of the initial-value problem is $\sqrt{h(t)} = (A_w\sqrt{H} - 4A_h t)/A_w$, where $A_w\sqrt{H} - 4A_h t \geq 0$. Thus, $h(t) = (A_w\sqrt{H} - 4A_h t)^2/A_w^2$ for $0 \leq t \leq A_w H/4A_h$.

(b) Identifying $H = 10$, $A_w = 4\pi$, and $A_h = \pi/576$ we have $h(t) = t^2/331{,}776 - (\sqrt{5/2}\,/144)t + 10$. Solving $h(t) = 0$ we see that the tank empties in $576\sqrt{10}$ seconds or 30.36 minutes.

12. When the height of the water is h, the radius of the top of the water is $\frac{2}{5}(20 - h)$ and $A_w = 4\pi(20 - h)^2/25$. The differential equation is

$$\frac{dh}{dt} = -c\frac{A_h}{A_w}\sqrt{2gh} = -0.6\frac{\pi(2/12)^2}{4\pi(20 - h)^2/25}\sqrt{64h} = -\frac{5}{6}\frac{\sqrt{h}}{(20 - h)^2}.$$

Separating variables and integrating we have

$$\frac{(20 - h)^2}{\sqrt{h}}\,dh = -\frac{5}{6}\,dt \quad \text{and} \quad 800\sqrt{h} - \frac{80}{3}h^{3/2} + \frac{2}{5}h^{5/2} = -\frac{5}{6}t + c.$$

Using $h(0) = 20$ we find $c = 2560\sqrt{5}/3$, so an implicit solution of the initial-value problem is

$$800\sqrt{h} - \frac{80}{3}h^{3/2} + \frac{2}{5}h^{5/2} = -\frac{5}{6}t + \frac{2560\sqrt{5}}{3}.$$

To find the time it takes the tank to empty we set $h = 0$ and solve for t. The tank empties in $1024\sqrt{5}$ seconds or 38.16 minutes. Thus, the tank empties more slowly when the base of the cone is on the bottom.

15. (a) Let ρ be the weight density of the water and V the volume of the object. Archimedes' principle states that the upward buoyant force has magnitude equal to the weight of the water displaced. Taking the positive direction to be down, the differential equation is

$$m\frac{dv}{dt} = mg - kv^2 - \rho V.$$

(b) Using separation of variables we have

$$\frac{m\,dv}{(mg-\rho V)-kv^2}=dt$$

$$\frac{m}{\sqrt{k}}\frac{\sqrt{k}\,dv}{(\sqrt{mg-\rho V})^2-(\sqrt{k}\,v)^2}=dt$$

$$\frac{m}{\sqrt{k}}\frac{1}{\sqrt{mg-\rho V}}\tanh^{-1}\frac{\sqrt{k}\,v}{\sqrt{mg-\rho V}}=t+c.$$

Thus

$$v(t)=\sqrt{\frac{mg-\rho V}{k}}\tanh\left(\frac{\sqrt{kmg-k\rho V}}{m}t+c_1\right).$$

(c) Since $\tanh t \to 1$ as $t \to \infty$, the terminal velocity is $\sqrt{(mg-\rho V)/k}$.

18. (a) Solving $r^2+(10-h)^2=10^2$ for r^2 we see that $r^2=20h-h^2$. Combining the rate of input of water, π, with the rate of output due to evaporation, $k\pi r^2=k\pi(20h-h^2)$, we have $dV/dt=\pi-k\pi(20h-h^2)$. Using $V=10\pi h^2-\frac{1}{3}\pi h^3$, we see also that $dV/dt=(20\pi h-\pi h^2)dh/dt$. Thus,

$$(20\pi h-\pi h^2)\frac{dh}{dt}=\pi-k\pi(20h-h^2)\quad\text{and}\quad\frac{dh}{dt}=\frac{1-20kh+kh^2}{20h-h^2}.$$

(b) Letting $k=1/100$, separating variables and integrating (with the help of a CAS), we get

$$\frac{100h(h-20)}{(h-10)^2}\,dh=dt\quad\text{and}\quad\frac{100(h^2-10h+100)}{10-h}=t+c.$$

Using $h(0)=0$ we find $c=1000$, and solving for h we get $h(t)=0.005(\sqrt{t^2+4000t}-t)$, where the positive square root is chosen because $h\geq 0$.

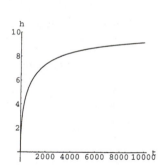

(c) The volume of the tank is $V=\frac{2}{3}\pi(10)^3$ feet, so at a rate of π cubic feet per minute, the tank will fill in $\frac{2}{3}(10)^3\approx 666.67$ minutes ≈ 11.11 hours.

(d) At 666.67 minutes, the depth of the water is $h(666.67)=5.486$ feet. From the graph in (b) we suspect that $\lim_{t\to\infty}h(t)=10$, in which case the tank will never completely fill. To prove this we compute the limit of $h(t)$:

$$\lim_{t\to\infty}h(t)=0.005\lim_{t\to\infty}\left(\sqrt{t^2+4000t}-t\right)=0.005\lim_{t\to\infty}\frac{t^2+4000t-t^2}{\sqrt{t^2+4000t}+t}$$

$$=0.005\lim_{t\to\infty}\frac{4000t}{t\sqrt{1+4000/t}+t}=0.005\lim_{t\to\infty}\frac{4000}{1+1}=0.005(2000)=10.$$

Exercises 3.3

3. The amounts of x and y are the same at about $t = 5$ days. The amounts of x and z are the same at about $t = 20$ days. The amounts of y and z are the same at about $t = 147$ days. The time when y and z are the same makes sense because most of A and half of B are gone, so half of C should have been formed.

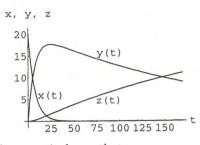

6. Let x_1, x_2, and x_3 be the amounts of salt in tanks A, B, and C, respectively, so that

$$x_1' = \frac{1}{100}x_2 \cdot 2 - \frac{1}{100}x_1 \cdot 6 = \frac{1}{50}x_2 - \frac{3}{50}x_1$$

$$x_2' = \frac{1}{100}x_1 \cdot 6 + \frac{1}{100}x_3 - \frac{1}{100}x_2 \cdot 2 - \frac{1}{100}x_2 \cdot 5 = \frac{3}{50}x_1 - \frac{7}{100}x_2 + \frac{1}{100}x_3$$

$$x_3' = \frac{1}{100}x_2 \cdot 5 - \frac{1}{100}x_3 - \frac{1}{100}x_3 \cdot 4 = \frac{1}{20}x_2 - \frac{1}{20}x_3.$$

9. From the graph we see that the populations are first equal at about $t = 5.6$. The approximate periods of x and y are both 45.

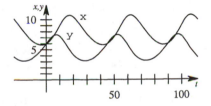

12. By Kirchoff's first law we have $i_1 = i_2 + i_3$. By Kirchoff's second law, on each loop we have $E(t) = Li_1' + R_1 i_2$ and $E(t) = Li_1' + R_2 i_3 + \frac{1}{C}q$ so that $q = CR_1 i_2 - CR_2 i_3$. Then $i_3 = q' = CR_1 i_2' - CR_2 i_3'$ so that the system is

$$Li_2' + Li_3' + R_1 i_2 = E(t)$$

$$-R_1 i_2' + R_2 i_3' + \frac{1}{C}i_3 = 0.$$

15. We first note that $s(t) + i(t) + r(t) = n$. Now the rate of change of the number of susceptible persons, $s(t)$, is proportional to the number of contacts between the number of people infected and the number who are susceptible; that is, $ds/dt = -k_1 s i$. We use $-k_1$ because $s(t)$ is decreasing. Next, the rate of change of the number of persons who have recovered is proportional to the number infected; that is, $dr/dt = k_2 i$ where k_2 is positive since r is increasing. Finally, to obtain di/dt we use

$$\frac{d}{dt}(s + i + r) = \frac{d}{dt}n = 0.$$

23

This gives

$$\frac{di}{dt} = -\frac{dr}{dt} - \frac{ds}{dt} = -k_2 i + k_1 s i.$$

The system of equations is then

$$\frac{ds}{dt} = -k_1 s i$$

$$\frac{di}{dt} = -k_2 i + k_1 s i$$

$$\frac{dr}{dt} = k_2 i.$$

A reasonable set of initial conditions is $i(0) = i_0$, the number of infected people at time 0, $s(0) = n - i_0$, and $r(0) = 0$.

Chapter 3 Review Exercises

3. (a) For $0 \le t < 4$, $6 \le t < 10$, and $12 \le t < 16$, no voltage is applied to the heart and $E(t) = 0$. At the other times the differential equation is $dE/dt = -E/RC$. Separating variables, integrating, and solving for E, we get $E = ke^{-t/RC}$, subject to $E(4) = E(10) = E(16) = 12$. These initial conditions yield, respectively, $k = 12e^{4/RC}$, $k = 12e^{10/RC}$, and $k = 12e^{16/RC}$. Thus

$$E(t) = \begin{cases} 0, & 0 \le t < 4, \ 6 \le t < 10, \ 12 \le t < 16 \\ 12e^{(4-t)/RC}, & 4 \le t < 6 \\ 12e^{(10-t)/RC}, & 10 \le t < 12 \\ 12e^{(16-t)/RC}, & 16 \le t < 18. \end{cases}$$

(b)

6. We first solve $\left(1 - \dfrac{t}{10}\right)\dfrac{di}{dt} + 0.2i = 4$. Separating variables we obtain

$$\frac{di}{40 - 2i} = \frac{dt}{10 - t}. \text{ Then}$$

$$-\frac{1}{2}\ln|40 - 2i| = -\ln|10 - t| + c \quad \text{or} \quad \sqrt{40 - 2i} = c_1(10 - t).$$

Since $i(0) = 0$ we must have $c_1 = 2/\sqrt{10}$. Solving for i we get $i(t) = 4t - \frac{1}{5}t^2$, $0 \le t < 10$. For $t \ge 10$ the equation for the current becomes $0.2i = 4$ or $i = 20$. Thus

$$i(t) = \begin{cases} 4t - \frac{1}{5}t^2, & 0 \le t < 10 \\ 20, & t \ge 10 \end{cases}.$$

9. From $\dfrac{dx}{dt} = k_1 x(\alpha - x)$ we obtain $\left(\dfrac{1/\alpha}{x} + \dfrac{1/\alpha}{\alpha - x} \right) dx = k_1 \, dt$ so that $x = \dfrac{\alpha c_1 e^{\alpha k_1 t}}{1 + c_1 e^{\alpha k_1 t}}$. From

$\dfrac{dy}{dt} = k_2 xy$ we obtain

$$\ln |y| = \frac{k_2}{k_1} \ln \left| 1 + c_1 e^{\alpha k_1 t} \right| + c \quad \text{or} \quad y = c_2 \left(1 + c_1 e^{\alpha k_1 t} \right)^{k_2/k_1}.$$

4 Higher-Order Differential Equations

Exercises 4.1

3. From $y = c_1 x + c_2 x \ln x$ we find $y' = c_1 + c_2(1 + \ln x)$. Then $y(1) = c_1 = 3$, $y'(1) = c_1 + c_2 = -1$ so that $c_1 = 3$ and $c_2 = -4$. The solution is $y = 3x - 4x \ln x$.

6. In this case we have $y(0) = c_1 = 0$, $y'(0) = 2c_2 \cdot 0 = 0$ so $c_1 = 0$ and c_2 is arbitrary. Two solutions are $y = x^2$ and $y = 2x^2$.

9. Since $a_2(x) = x - 2$ and $x_0 = 0$ the problem has a unique solution for $-\infty < x < 2$.

12. In this case we have $y(0) = c_1 = 1$, $y'(1) = 2c_2 = 6$ so that $c_1 = 1$ and $c_2 = 3$. The solution is $y = 1 + 3x^2$.

15. Since $(-4)x + (3)x^2 + (1)(4x - 3x^2) = 0$ the functions are linearly dependent.

18. Since $(1)\cos 2x + (1)1 + (-2)\cos^2 x = 0$ the functions are linearly dependent.

21. The functions are linearly independent since $W\left(1 + x, x, x^2\right) = \begin{vmatrix} 1+x & x & x^2 \\ 1 & 1 & 2x \\ 0 & 0 & 2 \end{vmatrix} = 2 \neq 0.$

24. The functions satisfy the differential equation and are linearly independent since

$$W(\cosh 2x, \sinh 2x) = 2$$

for $-\infty < x < \infty$. The general solution is

$$y = c_1 \cosh 2x + c_2 \sinh 2x.$$

27. The functions satisfy the differential equation and are linearly independent since

$$W\left(x^3, x^4\right) = x^6 \neq 0$$

for $0 < x < \infty$. The general solution is

$$y = c_1 x^3 + c_2 x^4.$$

30. The functions satisfy the differential equation and are linearly independent since

$$W(1, x, \cos x, \sin x) = 1$$

for $-\infty < x < \infty$. The general solution is

$$y = c_1 + c_2 x + c_3 \cos x + c_4 \sin x.$$

33. The functions $y_1 = e^{2x}$ and $y_2 = xe^{2x}$ form a fundamental set of solutions of the homogeneous equation, and $y_p = x^2 e^{2x} + x - 2$ is a particular solution of the nonhomogeneous equation.

36. (a) $y_{p_1} = 5$

(b) $y_{p_2} = -2x$

(c) $y_p = y_{p_1} + y_{p_2} = 5 - 2x$

(d) $y_p = \frac{1}{2}y_{p_1} - 2y_{p_2} = \frac{5}{2} + 4x$

_____ Exercises 4.2 _____

In Problems 3 and 6 we use reduction of order to find a second solution. In Problems 9-15 we use formula (5) from the text.

3. Define $y = u(x)\cos 4x$ so

$$y' = -4u\sin 4x + u'\cos 4x, \quad y'' = u''\cos 4x - 8u'\sin 4x - 16u\cos 4x$$

and

$$y'' + 16y = (\cos 4x)u'' - 8(\sin 4x)u' = 0 \quad \text{or} \quad u'' - 8(\tan 4x)u' = 0.$$

If $w = u'$ we obtain the first-order equation $w' - 8(\tan 4x)w = 0$ which has the integrating factor $e^{-8\int \tan 4x\, dx} = \cos^2 4x$. Now

$$\frac{d}{dx}[(\cos^2 4x)w] = 0 \quad \text{gives} \quad (\cos^2 4x)w = c.$$

Therefore $w = u' = c\sec^2 4x$ and $u = c_1\tan 4x$. A second solution is $y_2 = \tan 4x\cos 4x = \sin 4x$.

6. Define $y = u(x)e^{5x}$ so

$$y' = 5e^{5x}u + e^{5x}u', \quad y'' = e^{5x}u'' + 10e^{5x}u' + 25e^{5x}u$$

and

$$y'' - 25y = e^{5x}(u'' + 10u') = 0 \quad \text{or} \quad u'' + 10u' = 0.$$

If $w = u'$ we obtain the first-order equation $w' + 10w = 0$ which has the integrating factor $e^{10\int dx} = e^{10x}$. Now

$$\frac{d}{dx}[e^{10x}w] = 0 \quad \text{gives} \quad e^{10x}w = c.$$

Therefore $w = u' = ce^{-10x}$ and $u = c_1 e^{-10x}$. A second solution is $y_2 = e^{-10x}e^{5x} = e^{-5x}$.

9. Identifying $P(x) = -7/x$ we have

$$y_2 = x^4 \int \frac{e^{-\int -(7/x)\, dx}}{x^8}\, dx = x^4 \int \frac{1}{x}\, dx = x^4 \ln|x|.$$

27

A second solution is $y_2 = x^4 \ln |x|$.

12. Identifying $P(x) = 0$ we have

$$y_2 = x^{1/2} \ln x \int \frac{e^{-\int 0 \, dx}}{x (\ln x)^2} = x^{1/2} \ln x \left(-\frac{1}{\ln x} \right) = -x^{1/2}.$$

A second solution is $y_2 = x^{1/2}$.

15. Identifying $P(x) = 2(1 + x)/\left(1 - 2x - x^2\right)$ we have

$$y_2 = (x + 1) \int \frac{e^{-\int 2(1+x) \, dx /(1-2x-x^2)}}{(x+1)^2} \, dx = (x+1) \int \frac{e^{\ln\left(1-2x-x^2\right)}}{(x+1)^2} \, dx$$

$$= (x+1) \int \frac{1 - 2x - x^2}{(x+1)^2} \, dx = (x+1) \int \left[\frac{2}{(x+1)^2} - 1 \right] dx$$

$$= (x+1) \left[-\frac{2}{x+1} - x \right] = -2 - x^2 - x.$$

A second solution is $y_2 = x^2 + x + 2$.

18. Define $y = u(x) \cdot 1$ so

$$y' = u', \quad y'' = u'' \quad \text{and} \quad y'' + y' = u'' + u' = 0.$$

If $w = u'$ we obtain the first order equation $w' + w = 0$ which has the integrating factor $e^{\int dx} = e^x$. Now

$$\frac{d}{dx}[e^x w] = 0 \quad \text{gives} \quad e^x w = c.$$

Therefore $w = u' = ce^{-x}$ and $u = c_1 e^{-x}$. A second solution is $y_2 = 1 \cdot e^{-x} = e^{-x}$. We see by observation that a particular solution is $y_p = x$. The general solution is

$$y = c_1 + c_2 e^{-x} + x.$$

Exercises 4.3

3. From $m^2 - m - 6 = 0$ we obtain $m = 3$ and $m = -2$ so that $y = c_1 e^{3x} + c_2 e^{-2x}$.

6. From $m^2 - 10m + 25 = 0$ we obtain $m = 5$ and $m = 5$ so that $y = c_1 e^{5x} + c_2 x e^{5x}$.

9. From $m^2 + 9 = 0$ we obtain $m = 3i$ and $m = -3i$ so that $y = c_1 \cos 3x + c_2 \sin 3x$.

12. From $2m^2 + 2m + 1 = 0$ we obtain $m = -1/2 \pm i/2$ so that

$$y = e^{-x/2}(c_1 \cos x/2 + c_2 \sin x/2).$$

15. From $m^3 - 4m^2 - 5m = 0$ we obtain $m = 0$, $m = 5$, and $m = -1$ so that

$$y = c_1 + c_2 e^{5x} + c_3 e^{-x}.$$

18. From $m^3 + 3m^2 - 4m - 12 = 0$ we obtain $m = -2$, $m = 2$, and $m = -3$ so that

$$y = c_1 e^{-2x} + c_2 e^{2x} + c_3 e^{-3x}.$$

21. From $m^3 + 3m^2 + 3m + 1 = 0$ we obtain $m = -1$, $m = -1$, and $m = -1$ so that

$$y = c_1 e^{-x} + c_2 x e^{-x} + c_3 x^2 e^{-x}.$$

24. From $m^4 - 2m^2 + 1 = 0$ we obtain $m = 1$, $m = 1$, $m = -1$, and $m = -1$ so that

$$y = c_1 e^x + c_2 x e^x + c_3 e^{-x} + c_4 x e^{-x}.$$

27. From $m^5 + 5m^4 - 2m^3 - 10m^2 + m + 5 = 0$ we obtain $m = -1$, $m = -1$, $m = 1$, and $m = 1$, and $m = -5$ so that

$$u = c_1 e^{-r} + c_2 r e^{-r} + c_3 e^r + c_4 r e^r + c_5 e^{-5r}.$$

30. From $m^2 + 1 = 0$ we obtain $m = \pm i$ so that $y = c_1 \cos\theta + c_2 \sin\theta$. If $y(\pi/3) = 0$ and $y'(\pi/3) = 2$ then $\frac{1}{2}c_1 + \frac{\sqrt{3}}{2}c_2 = 0$, $-\frac{\sqrt{3}}{2}c_1 + \frac{1}{2}c_2 = 2$, so $c_1 = -\sqrt{3}$, $c_2 = 1$, and $y = -\sqrt{3}\cos\theta + \sin\theta$.

33. From $m^2 + m + 2 = 0$ we obtain $m = -1/2 \pm \sqrt{7}i/2$ so that $y = e^{-x/2}\left(c_1 \cos\sqrt{7}\,x/2 + c_2 \sin\sqrt{7}\,x/2\right)$. If $y(0) = 0$ and $y'(0) = 0$ then $c_1 = 0$ and $c_2 = 0$ so that $y = 0$.

36. From $m^3 + 2m^2 - 5m - 6 = 0$ we obtain $m = -1$, $m = 2$, and $m = -3$ so that

$$y = c_1 e^{-x} + c_2 e^{2x} + c_3 e^{-3x}.$$

If $y(0) = 0$, $y'(0) = 0$, and $y''(0) = 1$ then

$$c_1 + c_2 + c_3 = 0, \quad -c_1 + 2c_2 - 3c_3 = 0, \quad c_1 + 4c_2 + 9c_3 = 1,$$

so $c_1 = -1/6$, $c_2 = 1/15$, $c_3 = 1/10$, and

$$y = -\frac{1}{6}e^{-x} + \frac{1}{15}e^{2x} + \frac{1}{10}e^{-3x}.$$

39. From $m^2 + 1 = 0$ we obtain $m = \pm i$ so that $y = c_1 \cos x + c_2 \sin x$. If $y'(0) = 0$ and $y'(\pi/2) = 2$ then $c_1 = -2$, $c_2 = 0$, and $y = -2\cos x$.

42. The auxiliary equation is $m^2 - 1 = 0$ which has roots -1 and 1. By (10) the general solution is $y = c_1 e^x + c_2 e^{-x}$. By (11) the general solution is $y = c_1 \cosh x + c_2 \sinh x$. For $y = c_1 e^x + c_2 e^{-x}$ the boundary conditions imply $c_1 + c_2 = 1$, $c_1 e - c_2 e^{-1} = 0$. Solving for c_1 and c_2 we find $c_1 = 1/(1 + e^2)$ and $c_2 = e^2/(1 + e^2)$ so $y = e^x/(1 + e^2) + e^2 e^{-x}/(1 + e^2)$. For $y = c_1 \cosh x + c_2 \sinh x$ the boundary conditions imply $c_1 = 1$, $c_2 = -\tanh 1$, so $y = \cosh x - (\tanh 1)\sinh x$.

45. The auxiliary equation should have a pair of complex roots $a \pm bi$ where $a < 0$, so that the solution has the form $e^{ax}(c_1 \cos bx + c_2 \sin bx)$. Thus, the differential equation is (e).

48. The differential equation should have the form $y'' + k^2 y = 0$ where $k = 2$ so that the period of the solution is π. Thus, the differential equation is (b).

29

————————— **Exercises 4.4** —————————

3. From $m^2 - 10m + 25 = 0$ we find $m_1 = m_2 = 5$. Then $y_c = c_1 e^{5x} + c_2 x e^{5x}$ and we assume $y_p = Ax + B$. Substituting into the differential equation we obtain $25A = 30$ and $-10A + 25B = 3$. Then $A = \frac{6}{5}$, $B = \frac{6}{5}$, $y_p = \frac{6}{5}x + \frac{6}{5}$, and

$$y = c_1 e^{5x} + c_2 x e^{5x} + \frac{6}{5}x + \frac{6}{5}.$$

6. From $m^2 - 8m + 20 = 0$ we find $m_1 = 2 + 4i$ and $m_2 = 2 - 4i$. Then $y_c = e^{2x}(c_1 \cos 4x + c_2 \sin 4x)$ and we assume $y_p = Ax^2 + Bx + C + (Dx + E)e^x$. Substituting into the differential equation we obtain .

$$2A - 8B + 20C = 0$$

$$-6D + 13E = 0$$

$$-16A + 20B = 0$$

$$13D = -26$$

$$20A = 100.$$

Then $A = 5$, $B = 4$, $C = \frac{11}{10}$, $D = -2$, $E = -\frac{12}{13}$, $y_p = 5x^2 + 4x + \frac{11}{10} + \left(-2x - \frac{12}{13}\right)e^x$ and

$$y = e^{2x}(c_1 \cos 4x + c_2 \sin 4x) + 5x^2 + 4x + \frac{11}{10} + \left(-2x - \frac{12}{13}\right)e^x.$$

9. From $m^2 - m = 0$ we find $m_1 = 1$ and $m_2 = 0$. Then $y_c = c_1 e^x + c_2$ and we assume $y_p = Ax$. Substituting into the differential equation we obtain $-A = -3$. Then $A = 3$, $y_p = 3x$ and $y = c_1 e^x + c_2 + 3x$.

12. From $m^2 - 16 = 0$ we find $m_1 = 4$ and $m_2 = -4$. Then $y_c = c_1 e^{4x} + c_2 e^{-4x}$ and we assume $y_p = Axe^{4x}$. Substituting into the differential equation we obtain $8A = 2$. Then $A = \frac{1}{4}$, $y_p = \frac{1}{4}xe^{4x}$ and

$$y = c_1 e^{4x} + c_2 e^{-4x} + \frac{1}{4}xe^{4x}.$$

15. From $m^2 + 1 = 0$ we find $m_1 = i$ and $m_2 = -i$. Then $y_c = c_1 \cos x + c_2 \sin x$ and we assume $y_p = (Ax^2 + Bx)\cos x + (Cx^2 + Dx)\sin x$. Substituting into the differential equation we obtain $4C = 0$, $2A + 2D = 0$, $-4A = 2$, and $-2B + 2C = 0$. Then $A = -\frac{1}{2}$, $B = 0$, $C = 0$, $D = \frac{1}{2}$, $y_p = -\frac{1}{2}x^2 \cos x + \frac{1}{2}x \sin x$, and

$$y = c_1 \cos x + c_2 \sin x - \frac{1}{2}x^2 \cos x + \frac{1}{2}x \sin x.$$

30

18. From $m^2 - 2m + 2 = 0$ we find $m_1 = 1 + i$ and $m_2 = 1 - i$. Then $y_c = e^x(c_1 \cos x + c_2 \sin x)$ and we assume $y_p = Ae^{2x} \cos x + Be^{2x} \sin x$. Substituting into the differential equation we obtain $A + 2B = 1$ and $-2A + B = -3$. Then $A = \frac{7}{5}$, $B = -\frac{1}{5}$, $y_p = \frac{7}{5}e^{2x} \cos x - \frac{1}{5}e^{2x} \sin x$ and

$$y = e^x(c_1 \cos x + c_2 \sin x) + \frac{7}{5}e^{2x} \cos x - \frac{1}{5}e^{2x} \sin x.$$

21. From $m^3 - 6m^2 = 0$ we find $m_1 = m_2 = 0$ and $m_3 = 6$. Then $y_c = c_1 + c_2 x + c_3 e^{6x}$ and we assume $y_p = Ax^2 + B \cos x + C \sin x$. Substituting into the differential equation we obtain $-12A = 3$, $6B - C = -1$, and $B + 6C = 0$. Then $A = -\frac{1}{4}$, $B = -\frac{6}{37}$, $C = \frac{1}{37}$, $y_p = -\frac{1}{4}x^2 - \frac{6}{37} \cos x + \frac{1}{37} \sin x$, and

$$y = c_1 + c_2 x + c_3 e^{6x} - \frac{1}{4}x^2 - \frac{6}{37} \cos x + \frac{1}{37} \sin x.$$

24. From $m^3 - m^2 - 4m + 4 = 0$ we find $m_1 = 1$, $m_2 = 2$, and $m_3 = -2$. Then $y_c = c_1 e^x + c_2 e^{2x} + c_3 e^{-2x}$ and we assume $y_p = A + Bxe^x + Cxe^{2x}$. Substituting into the differential equation we obtain $4A = 5$, $-3B = -1$, and $4C = 1$. Then $A = \frac{5}{4}$, $B = \frac{1}{3}$, $C = \frac{1}{4}$, $y_p = \frac{5}{4} + \frac{1}{3}xe^x + \frac{1}{4}xe^{2x}$, and

$$y = c_1 e^x + c_2 e^{2x} + c_3 e^{-2x} + \frac{5}{4} + \frac{1}{3}xe^x + \frac{1}{4}xe^{2x}.$$

27. We have $y_c = c_1 \cos 2x + c_2 \sin 2x$ and we assume $y_p = A$. Substituting into the differential equation we find $A = -\frac{1}{2}$. Thus $y = c_1 \cos 2x + c_2 \sin 2x - \frac{1}{2}$. From the initial conditions we obtain $c_1 = 0$ and $c_2 = \sqrt{2}$, so $y = \sqrt{2} \sin 2x - \frac{1}{2}$.

30. We have $y_c = c_1 e^{-2x} + c_2 xe^{-2x}$ and we assume $y_p = (Ax^3 + Bx^2)e^{-2x}$. Substituting into the differential equation we find $A = \frac{1}{6}$ and $B = \frac{3}{2}$. Thus $y = c_1 e^{-2x} + c_2 xe^{-2x} + \left(\frac{1}{6}x^3 + \frac{3}{2}x^2\right)e^{-2x}$. From the initial conditions we obtain $c_1 = 2$ and $c_2 = 9$, so

$$y = 2e^{-2x} + 9xe^{-2x} + \left(\frac{1}{6}x^3 + \frac{3}{2}x^2\right)e^{-2x}.$$

33. We have $x_c = c_1 \cos \omega t + c_2 \sin \omega t$ and we assume $x_p = At \cos \omega t + Bt \sin \omega t$. Substituting into the differential equation we find $A = -F_0/2\omega$ and $B = 0$. Thus $x = c_1 \cos \omega t + c_2 \sin \omega t - (F_0/2\omega)t \cos \omega t$. From the initial conditions we obtain $c_1 = 0$ and $c_2 = F_0/2\omega^2$, so

$$x = (F_0/2\omega^2) \sin \omega t - (F_0/2\omega)t \cos \omega t.$$

36. We have $y_c = c_1 e^{-2x} + e^x(c_2 \cos \sqrt{3}\,x + c_3 \sin \sqrt{3}\,x)$ and we assume $y_p = Ax + B + Cxe^{-2x}$. Substituting into the differential equation we find $A = \frac{1}{4}$, $B = -\frac{5}{8}$, and $C = \frac{2}{3}$. Thus

$$y = c_1 e^{-2x} + e^x(c_2 \cos \sqrt{3}\,x + c_3 \sin \sqrt{3}\,x) + \frac{1}{4}x - \frac{5}{8} + \frac{2}{3}xe^{-2x}.$$

From the initial conditions we obtain $c_1 = -\frac{23}{12}$, $c_2 = -\frac{59}{24}$, and $c_3 = \frac{17}{72}\sqrt{3}$, so

$$y = -\frac{23}{12}e^{-2x} + e^x\left(-\frac{59}{24}\cos\sqrt{3}\,x + \frac{17}{72}\sqrt{3}\sin\sqrt{3}\,x\right) + \frac{1}{4}x - \frac{5}{8} + \frac{2}{3}xe^{-2x}.$$

39. We have $y_c = c_1 \cos 2x + c_2 \sin 2x$ and we assume $y_p = A \cos x + B \sin x$ on $[0, \pi/2]$. Substituting into the differential equation we find $A = 0$ and $B = \frac{1}{3}$. Thus $y = c_1 \cos 2x + c_2 \sin 2x + \frac{1}{3} \sin x$ on $[0, \pi/2]$. On $(\pi/2, \infty)$ we have $y = c_3 \cos 2x + c_4 \sin 2x$. From $y(0) = 1$ and $y'(0) = 2$ we obtain

$$c_1 = 1$$

$$\frac{1}{3} + 2c_2 = 2.$$

Solving this system we find $c_1 = 1$ and $c_2 = \frac{5}{6}$. Thus $y = \cos 2x + \frac{5}{6} \sin 2x + \frac{1}{3} \sin x$ on $[0, \pi/2]$. Now continuity of y at $x = \pi/2$ implies

$$\cos \pi + \frac{5}{6} \sin \pi + \frac{1}{3} \sin \frac{\pi}{2} = c_3 \cos \pi + c_4 \sin \pi$$

or $-1 + \frac{1}{3} = -c_3$. Hence $c_3 = \frac{2}{3}$. Continuity of y' at $x = \pi/2$ implies

$$-2 \sin \pi + \frac{5}{3} \cos \pi + \frac{1}{3} \cos \frac{\pi}{2} = -2c_3 \sin \pi + 2c_4 \cos \pi$$

or $-\frac{5}{3} = -2c_4$. Then $c_4 = \frac{5}{6}$ and the solution of the initial-value problem is

$$y(x) = \begin{cases} \cos 2x + \frac{5}{6} \sin 2x + \frac{1}{3} \sin x, & 0 \le x \le \pi/2 \\ \frac{2}{3} \cos 2x + \frac{5}{6} \sin 2x, & x > \pi/2. \end{cases}$$

Exercises 4.5

3. $(D^2 - 4D - 12)y = (D - 6)(D + 2)y = x - 6$

6. $(D^3 + 4D)y = D(D^2 + 4)y = e^x \cos 2x$

9. $(D^4 + 8D)y = D(D + 2)(D^2 - 2D + 4)y = 4$

12. $(2D - 1)y = (2D - 1)4e^{x/2} = 8De^{x/2} - 4e^{x/2} = 4e^{x/2} - 4e^{x/2} = 0$

15. D^4 because of x^3

18. $D^2(D - 6)^2$ because of x and xe^{6x}

21. $D^3(D^2 + 16)$ because of x^2 and $\sin 4x$

24. $D(D - 1)(D - 2)$ because of 1, e^x, and e^{2x}

27. $1, x, x^2, x^3, x^4$

30. $D^2 - 9D - 36 = (D - 12)(D + 3)$; e^{12x}, e^{-3x}

33. $D^3 - 10D^2 + 25D = D(D - 5)^2$; $1, e^{5x}, xe^{5x}$

36. Applying D to the differential equation we obtain

$$D(2D^2 - 7D + 5)y = 0.$$

Then

$$y = \underbrace{c_1 e^{5x/2} + c_2 e^x}_{y_c} + c_3$$

and $y_p = A$. Substituting y_p into the differential equation yields $5A = -29$ or $A = -29/5$. The general solution is

$$y = c_1 e^{5x/2} + c_2 e^x - \frac{29}{5}.$$

39. Applying D^2 to the differential equation we obtain

$$D^2(D^2 + 4D + 4)y = D^2(D+2)^2 y = 0.$$

Then

$$y = \underbrace{c_1 e^{-2x} + c_2 x e^{-2x}}_{y_c} + c_3 + c_4 x$$

and $y_p = Ax + B$. Substituting y_p into the differential equation yields $4Ax + (4A + 4B) = 2x + 6$. Equating coefficients gives

$$4A = 2$$

$$4A + 4B = 6.$$

Then $A = 1/2$, $B = 1$, and the general solution is

$$y = c_1 e^{-2x} + c_2 x e^{-2x} + \frac{1}{2}x + 1.$$

42. Applying D^4 to the differential equation we obtain

$$D^4(D^2 - 2D + 1)y = D^4(D-1)^2 y = 0.$$

Then

$$y = \underbrace{c_1 e^x + c_2 x e^x}_{y_c} + c_3 x^3 + c_4 x^2 + c_5 x + c_6$$

and $y_p = Ax^3 + Bx^2 + Cx + D$. Substituting y_p into the differential equation yields

$$Ax^3 + (B - 6A)x^2 + (6A - 4B + C)x + (2B - 2C + D) = x^3 + 4x.$$

Equating coefficients gives

$$A = 1$$

$$B - 6A = 0$$

$$6A - 4B + C = 4$$

$$2B - 2C + D = 0.$$

33

Then $A = 1$, $B = 6$, $C = 22$, $D = 32$, and the general solution is

$$y = c_1 e^x + c_2 x e^x + x^3 + 6x^2 + 22x + 32.$$

45. Applying $D(D-1)$ to the differential equation we obtain

$$D(D-1)(D^2 - 2D - 3)y = D(D-1)(D+1)(D-3)y = 0.$$

Then

$$y = \underbrace{c_1 e^{3x} + c_2 e^{-x}}_{y_c} + c_3 e^x + c_4$$

and $y_p = Ae^x + B$. Substituting y_p into the differential equation yields $-4Ae^x - 3B = 4e^x - 9$. Equating coefficients gives $A = -1$ and $B = 3$. The general solution is

$$y = c_1 e^{3x} + c_2 e^{-x} - e^x + 3.$$

48. Applying $D(D^2 + 1)$ to the differential equation we obtain

$$D(D^2 + 1)(D^2 + 4)y = 0.$$

Then

$$y = \underbrace{c_1 \cos 2x + c_2 \sin 2x}_{y_c} + c_3 \cos x + c_4 \sin x + c_5$$

and $y_p = A \cos x + B \sin x + C$. Substituting y_p into the differential equation yields

$$3A \cos x + 3B \sin x + 4C = 4 \cos x + 3 \sin x - 8.$$

Equating coefficients gives $A = 4/3$, $B = 1$, and $C = -2$. The general solution is

$$y = c_1 \cos 2x + c_2 \sin 2x + \frac{4}{3} \cos x + \sin x - 2.$$

51. Applying $D(D-1)^3$ to the differential equation we obtain

$$D(D-1)^3(D^2 - 1)y = D(D-1)^4(D+1)y = 0.$$

Then

$$y = \underbrace{c_1 e^x + c_2 e^{-x}}_{y_c} + c_3 x^3 e^x + c_4 x^2 e^x + c_5 x e^x + c_6$$

and $y_p = Ax^3 e^x + Bx^2 e^x + Cxe^x + D$. Substituting y_p into the differential equation yields

$$6Ax^2 e^x + (6A + 4B)xe^x + (2B + 2C)e^x - D = x^2 e^x + 5.$$

Equating coefficients gives

$$6A = 1$$

$$6A + 4B = 0$$

$$2B + 2C = 0$$

$$-D = 5.$$

Then $A = 1/6$, $B = -1/4$, $C = 1/4$, $D = -5$, and the general solution is

$$y = c_1 e^x + c_2 e^{-x} + \frac{1}{6}x^3 e^x - \frac{1}{4}x^2 e^x + \frac{1}{4}xe^x - 5.$$

54. Applying $D^2 - 2D + 10$ to the differential equation we obtain

$$(D^2 - 2D + 10)\left(D^2 + D + \frac{1}{4}\right)y = (D^2 - 2D + 10)\left(D + \frac{1}{2}\right)^2 y = 0.$$

Then

$$y = \underbrace{c_1 e^{-x/2} + c_2 x e^{-x/2}}_{y_c} + c_3 e^x \cos 3x + c_4 e^x \sin 3x$$

and $y_p = Ae^x \cos 3x + Be^x \sin 3x$. Substituting y_p into the differential equation yields

$$(9B - 27A/4)e^x \cos 3x - (9A + 27B/4)e^x \sin 3x = -e^x \cos 3x + e^x \sin 3x.$$

Equating coefficients gives

$$-\frac{27}{4}A + 9B = -1$$

$$-9A - \frac{27}{4}B = 1.$$

Then $A = -4/225$, $B = -28/225$, and the general solution is

$$y = c_1 e^{-x/2} + c_2 x e^{-x/2} - \frac{4}{225}e^x \cos 3x - \frac{28}{225}e^x \sin 3x.$$

57. Applying $(D^2 + 1)^2$ to the differential equation we obtain

$$(D^2 + 1)^2 (D^2 + D + 1) = 0.$$

Then

$$y = \underbrace{e^{-x/2}\left[c_1 \cos \frac{\sqrt{3}}{2}x + c_2 \sin \frac{\sqrt{3}}{2}x\right]}_{y_c} + c_3 \cos x + c_4 \sin x + c_5 x \cos x + c_6 x \sin x$$

and $y_p = A \cos x + B \sin x + Cx \cos x + Dx \sin x$. Substituting y_p into the differential equation yields

$$(B + C + 2D) \cos x + Dx \cos x + (-A - 2C + D) \sin x - Cx \sin x = x \sin x.$$

Equating coefficients gives

$$B + C + 2D = 0$$

$$D = 0$$

$$-A - 2C + D = 0$$

$$-C = 1.$$

35

Then $A = 2$, $B = 1$, $C = -1$, and $D = 0$, and the general solution is

$$y = e^{-x/2} \left[c_1 \cos \frac{\sqrt{3}}{2} x + c_2 \sin \frac{\sqrt{3}}{2} x \right] + 2 \cos x + \sin x - x \cos x.$$

60. Applying $D(D-1)^2(D+1)$ to the differential equation we obtain

$$D(D-1)^2(D+1)(D^3 - D^2 + D - 1) = D(D-1)^3(D+1)(D^2+1) = 0.$$

Then

$$y = \underbrace{c_1 e^x + c_2 \cos x + c_3 \sin x}_{y_c} + c_4 + c_5 e^{-x} + c_6 x e^x + c_7 x^2 e^x$$

and $y_p = A + Be^{-x} + Cxe^x + Dx^2 e^x$. Substituting y_p into the differential equation yields

$$4Dxe^x + (2C + 4D)e^x - 4Be^{-x} - A = xe^x - e^{-x} + 7.$$

Equating coefficients gives

$$4D = 1$$

$$2C + 4D = 0$$

$$-4B = -1$$

$$-A = 7.$$

Then $A = -7$, $B = 1/4$, $C = -1/2$, and $D = 1/4$, and the general solution is

$$y = c_1 e^x + c_2 \cos x + c_3 \sin x - 7 + \frac{1}{4}e^{-x} - \frac{1}{2}xe^x + \frac{1}{4}x^2 e^x.$$

63. Applying $D(D-1)$ to the differential equation we obtain

$$D(D-1)(D^4 - 2D^3 + D^2) = D^3(D-1)^3 = 0.$$

Then

$$y = \underbrace{c_1 + c_2 x + c_3 e^x + c_4 x e^x}_{y_c} + c_5 x^2 + c_6 x^2 e^x$$

and $y_p = Ax^2 + Bx^2 e^x$. Substituting y_p into the differential equation yields $2A + 2Be^x = 1 + e^x$. Equating coefficients gives $A = 1/2$ and $B = 1/2$. The general solution is

$$y = c_1 + c_2 x + c_3 e^x + c_4 x e^x + \frac{1}{2}x^2 + \frac{1}{2}x^2 e^x.$$

66. The complementary function is $y_c = c_1 + c_2 e^{-x}$. Using D^2 to annihilate x we find $y_p = Ax + Bx^2$. Substituting y_p into the differential equation we obtain $(A + 2B) + 2Bx = x$. Thus $A = -1$ and $B = 1/2$, and

$$y = c_1 + c_2 e^{-x} - x + \frac{1}{2}x^2$$

$$y' = -c_2 e^{-x} - 1 + x.$$

The initial conditions imply

$$c_1 + c_2 = 1$$

$$-c_2 = 1.$$

Thus $c_1 = 2$ and $c_2 = -1$, and

$$y = 2 - e^{-x} - x + \frac{1}{2}x^2.$$

69. The complementary function is $y_c = c_1 \cos x + c_2 \sin x$. Using $(D^2 + 1)(D^2 + 4)$ to annihilate $8 \cos 2x - 4 \sin x$ we find $y_p = Ax \cos x + Bx \sin x + C \cos 2x + D \sin 2x$. Substituting y_p into the differential equation we obtain $2B \cos x - 3C \cos 2x - 2A \sin x - 3D \sin 2x = 8 \cos 2x - 4 \sin x$. Thus $A = 2$, $B = 0$, $C = -8/3$, and $D = 0$, and

$$y = c_1 \cos x + c_2 \sin x + 2x \cos x - \frac{8}{3} \cos 2x$$

$$y' = -c_1 \sin x + c_2 \cos x + 2 \cos x - 2x \sin x + \frac{16}{3} \sin 2x.$$

The initial conditions imply

$$c_2 + \frac{8}{3} = -1$$

$$-c_1 - \pi = 0.$$

Thus $c_1 = -\pi$ and $c_2 = -11/3$, and

$$y = -\pi \cos x - \frac{11}{3} \sin x + 2x \cos x - \frac{8}{3} \cos 2x.$$

72. The complementary function is $y_c = c_1 + c_2 x + c_3 x^2 + c_4 e^x$. Using $D^2(D - 1)$ to annihilate $x + e^x$ we find $y_p = Ax^3 + Bx^4 + Cxe^x$. Substituting y_p into the differential equation we obtain $(-6A + 24B) - 24Bx + Ce^x = x + e^x$. Thus $A = -1/6$, $B = -1/24$, and $C = 1$, and

$$y = c_1 + c_2 x + c_3 x^2 + c_4 e^x - \frac{1}{6}x^3 - \frac{1}{24}x^4 + xe^x$$

$$y' = c_2 + 2c_3 x + c_4 e^x - \frac{1}{2}x^2 - \frac{1}{6}x^3 + e^x + xe^x$$

$$y'' = 2c_3 + c_4 e^x - x - \frac{1}{2}x^2 + 2e^x + xe^x.$$

$$y''' = c_4 e^x - 1 - x + 3e^x + xe^x$$

37

The initial conditions imply

$$c_1 + c_4 = 0$$

$$c_2 + c_4 + 1 = 0$$

$$2c_3 + c_4 + 2 = 0$$

$$2 + c_4 = 0.$$

Thus $c_1 = 2$, $c_2 = 1$, $c_3 = 0$, and $c_4 = -2$, and

$$y = 2 + x - 2e^x - \frac{1}{6}x^3 - \frac{1}{24}x^4 + xe^x.$$

Exercises 4.6

The particular solution, $y_p = u_1 y_1 + u_2 y_2$, in the following problems can take on a variety of forms, especially where trigonometric functions are involved. The validity of a particular form can best be checked by substituting it back into the differential equation.

3. The auxiliary equation is $m^2 + 1 = 0$, so $y_c = c_1 \cos x + c_2 \sin x$ and

$$W = \begin{vmatrix} \cos x & \sin x \\ -\sin x & \cos x \end{vmatrix} = 1.$$

Identifying $f(x) = \sin x$ we obtain

$$u_1' = -\sin^2 x$$

$$u_2' = \cos x \sin x.$$

Then

$$u_1 = \frac{1}{4}\sin 2x - \frac{1}{2}x = \frac{1}{2}\sin x \cos x - \frac{1}{2}x$$

$$u_2 = -\frac{1}{2}\cos^2 x.$$

and

$$y = c_1 \cos x + c_2 \sin x + \frac{1}{2}\sin x \cos^2 x - \frac{1}{2}x \cos x - \frac{1}{2}\cos^2 x \sin x$$

$$= c_1 \cos x + c_2 \sin x - \frac{1}{2}x \cos x.$$

6. The auxiliary equation is $m^2 + 1 = 0$, so $y_c = c_1 \cos x + c_2 \sin x$ and

$$W = \begin{vmatrix} \cos x & \sin x \\ -\sin x & \cos x \end{vmatrix} = 1.$$

Identifying $f(x) = \sec^2 x$ we obtain

$$u_1' = -\frac{\sin x}{\cos^2 x}$$

$$u_2' = \sec x.$$

Then

$$u_1 = -\frac{1}{\cos x} = -\sec x$$

$$u_2 = \ln|\sec x + \tan x|$$

and

$$y = c_1 \cos x + c_2 \sin x - \cos x \sec x + \sin x \ln|\sec x + \tan x|$$

$$= c_1 \cos x + c_2 \sin x - 1 + \sin x \ln|\sec x + \tan x|.$$

9. The auxiliary equation is $m^2 - 4 = 0$, so $y_c = c_1 e^{2x} + c_2 e^{-2x}$ and

$$W = \begin{vmatrix} e^{2x} & e^{-2x} \\ 2e^{2x} & -2e^{-2x} \end{vmatrix} = -4.$$

Identifying $f(x) = e^{2x}/x$ we obtain $u_1' = 1/4x$ and $u_2' = -e^{4x}/4x$. Then

$$u_1 = \frac{1}{4}\ln|x|,$$

$$u_2 = -\frac{1}{4}\int_{x_0}^x \frac{e^{4t}}{t}\,dt$$

and

$$y = c_1 e^{2x} + c_2 e^{-2x} + \frac{1}{4}\left(e^{2x}\ln|x| - e^{-2x}\int_{x_0}^x \frac{e^{4t}}{t}\,dt\right), \qquad x_0 > 0.$$

12. The auxiliary equation is $m^2 - 2m + 1 = (m-1)^2 = 0$, so $y_c = c_1 e^x + c_2 x e^x$ and

$$W = \begin{vmatrix} e^x & xe^x \\ e^x & xe^x + e^x \end{vmatrix} = e^{2x}.$$

Identifying $f(x) = e^x/\left(1 + x^2\right)$ we obtain

$$u_1' = -\frac{xe^x e^x}{e^{2x}\left(1 + x^2\right)} = -\frac{x}{1 + x^2}$$

$$u_2' = \frac{e^x e^x}{e^{2x}\left(1 + x^2\right)} = \frac{1}{1 + x^2}.$$

Then $u_1 = -\frac{1}{2}\ln\left(1 + x^2\right)$, $u_2 = \tan^{-1} x$, and

$$y = c_1 e^x + c_2 x e^x - \frac{1}{2}e^x \ln\left(1 + x^2\right) + xe^x \tan^{-1} x.$$

15. The auxiliary equation is $m^2 + 2m + 1 = (m+1)^2 = 0$, so $y_c = c_1 e^{-t} + c_2 t e^{-t}$ and

$$W = \begin{vmatrix} e^{-t} & t e^{-t} \\ -e^{-t} & -t e^{-t} + e^{-t} \end{vmatrix} = e^{-2t}.$$

Identifying $f(t) = e^{-t} \ln t$ we obtain

$$u_1' = -\frac{t e^{-t} e^{-t} \ln t}{e^{-2t}} = -t \ln t$$

$$u_2' = \frac{e^{-t} e^{-t} \ln t}{e^{-2t}} = \ln t.$$

Then

$$u_1 = -\frac{1}{2} t^2 \ln t + \frac{1}{4} t^2$$

$$u_2 = t \ln t - t$$

and

$$y = c_1 e^{-t} + c_2 t e^{-t} - \frac{1}{2} t^2 e^{-t} \ln t + \frac{1}{4} t^2 e^{-t} + t^2 e^{-t} \ln t - t^2 e^{-t}$$

$$= c_1 e^{-t} + c_2 t e^{-t} + \frac{1}{2} t^2 e^{-t} \ln t - \frac{3}{4} t^2 e^{-t}.$$

18. The auxiliary equation is $4m^2 - 4m + 1 = (2m - 1)^2 = 0$, so $y_c = c_1 e^{x/2} + c_2 x e^{x/2}$ and

$$W = \begin{vmatrix} e^{x/2} & x e^{x/2} \\ \frac{1}{2} e^{x/2} & \frac{1}{2} x e^{x/2} + e^{x/2} \end{vmatrix} = e^x.$$

Identifying $f(x) = \frac{1}{4} e^{x/2} \sqrt{1 - x^2}$ we obtain

$$u_1' = -\frac{x e^{x/2} e^{x/2} \sqrt{1 - x^2}}{4 e^x} = -\frac{1}{4} x \sqrt{1 - x^2}$$

$$u_2' = \frac{e^{x/2} e^{x/2} \sqrt{1 - x^2}}{4 e^x} = \frac{1}{4} \sqrt{1 - x^2}.$$

Then

$$u_1 = \frac{1}{12} \left(1 - x^2\right)^{3/2}$$

$$u_2 = \frac{x}{8} \sqrt{1 - x^2} + \frac{1}{8} \sin^{-1} x$$

and

$$y = c_1 e^{x/2} + c_2 x e^{x/2} + \frac{1}{12} e^{x/2} \left(1 - x^2\right)^{3/2} + \frac{1}{8} x^2 e^{x/2} \sqrt{1 - x^2} + \frac{1}{8} x e^{x/2} \sin^{-1} x.$$

21. The auxiliary equation is $m^2 + 2m - 8 = (m - 2)(m + 4) = 0$, so $y_c = c_1 e^{2x} + c_2 e^{-4x}$ and

$$W = \begin{vmatrix} e^{2x} & e^{-4x} \\ 2e^{2x} & -4e^{-4x} \end{vmatrix} = -6 e^{-2x}.$$

40

Identifying $f(x) = 2e^{-2x} - e^{-x}$ we obtain

$$u_1' = \frac{1}{3}e^{-4x} - \frac{1}{6}e^{-3x}$$

$$u_2' = -\frac{1}{6}e^{3x} - \frac{1}{3}e^{2x}.$$

Then

$$u_1 = -\frac{1}{12}e^{-4x} + \frac{1}{18}e^{-3x}$$

$$u_2 = \frac{1}{18}e^{3x} - \frac{1}{6}e^{2x}.$$

Thus

$$y = c_1 e^{2x} + c_2 e^{-4x} - \frac{1}{12}e^{-2x} + \frac{1}{18}e^{-x} + \frac{1}{18}e^{-x} - \frac{1}{6}e^{-2x}$$

$$= c_1 e^{2x} + c_2 e^{-4x} - \frac{1}{4}e^{-2x} + \frac{1}{9}e^{-x}$$

and

$$y' = 2c_1 e^{2x} - 4c_2 e^{-4x} + \frac{1}{2}e^{-2x} - \frac{1}{9}e^{-x}.$$

The initial conditions imply

$$c_1 + c_2 - \frac{5}{36} = 1$$

$$2c_1 - 4c_2 + \frac{7}{18} = 0.$$

Thus $c_1 = 25/36$ and $c_2 = 4/9$, and

$$y = \frac{25}{36}e^{2x} + \frac{4}{9}e^{-4x} - \frac{1}{4}e^{-2x} + \frac{1}{9}e^{-x}.$$

24. Write the equation in the form

$$y'' + \frac{1}{x}y' + \frac{1}{x^2}y = \frac{\sec(\ln x)}{x^2}$$

and identify $f(x) = \sec(\ln x)/x^2$. From $y_1 = \cos(\ln x)$ and $y_2 = \sin(\ln x)$ we compute

$$W = \begin{vmatrix} \cos(\ln x) & \sin(\ln x) \\ -\dfrac{\sin(\ln x)}{x} & \dfrac{\cos(\ln x)}{x} \end{vmatrix} = \frac{1}{x}.$$

Now

$$u_1' = -\frac{\tan(\ln x)}{x} \quad \text{so} \quad u_1 = \ln|\cos(\ln x)|,$$

and

$$u_2' = \frac{1}{x} \quad \text{so} \quad u_2 = \ln x.$$

41

Thus, a particular solution is

$$y_p = \cos(\ln x) \ln |\cos(\ln x)| + (\ln x) \sin(\ln x).$$

Exercises 4.7

3. The auxiliary equation is $m^2 = 0$ so that $y = c_1 + c_2 \ln x$.

6. The auxiliary equation is $m^2 + 4m + 3 = (m+1)(m+3) = 0$ so that $y = c_1 x^{-1} + c_2 x^{-3}$.

9. The auxiliary equation is $25m^2 + 1 = 0$ so that $y = c_1 \cos\left(\frac{1}{5}\ln x\right) + c_2 \left(\frac{1}{5}\ln x\right)$.

12. The auxiliary equation is $m^2 + 7m + 6 = (m+1)(m+6) = 0$ so that $y = c_1 x^{-1} + c_2 x^{-6}$.

15. Assuming that $y = x^m$ and substituting into the differential equation we obtain

$$m(m-1)(m-2) - 6 = m^3 - 3m^2 + 2m - 6 = (m-3)(m^2+2) = 0.$$

Thus

$$y = c_1 x^3 + c_2 \cos\left(\sqrt{2}\ln x\right) + c_3 \sin\left(\sqrt{2}\ln x\right).$$

18. Assuming that $y = x^m$ and substituting into the differential equation we obtain

$$m(m-1)(m-2)(m-3) + 6m(m-1)(m-2) + 9m(m-1) + 3m + 1 = m^4 + 2m^2 + 1 = (m^2+1)^2 = 0.$$

Thus

$$y = c_1 \cos(\ln x) + c_2 \sin(\ln x) + c_3 \ln x \cos(\ln x) + c_4 \ln x \sin(\ln x).$$

21. The auxiliary equation is $m^2 - 2m + 1 = (m-1)^2 = 0$ so that $y_c = c_1 x + c_2 x \ln x$ and

$$W(x, x\ln x) = \begin{vmatrix} x & x\ln x \\ 1 & 1 + \ln x \end{vmatrix} = x.$$

Identifying $f(x) = 2/x$ we obtain $u_1' = -2\ln x/x$ and $u_2' = 2/x$. Then $u_1 = -(\ln x)^2$, $u_2 = 2\ln x$, and

$$y = c_1 x + c_2 x \ln x - x(\ln x)^2 + 2x(\ln x)^2$$

$$= c_1 x + c_2 x \ln x + x(\ln x)^2.$$

24. The auxiliary equation is $m^2 - 6m + 8 = (m-2)(m-4) = 0$, so that

$$y = c_1 x^2 + c_2 x^4 \quad \text{and} \quad y' = 2c_1 x + 4c_2 x^3.$$

The initial conditions imply

$$4c_1 + 16c_2 = 32$$

$$4c_1 + 32c_2 = 0.$$

Thus, $c_1 = 16$, $c_2 = -2$, and $y = 16x^2 - 2x^4$.

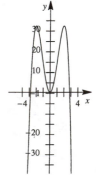

27. The auxiliary equation is $m^2 = 0$ so that $y_c = c_1 + c_2 \ln x$ and

$$W(1, \ln x) = \begin{vmatrix} 1 & \ln x \\ 0 & 1/x \end{vmatrix} = \frac{1}{x}.$$

Identifying $f(x) = 1$ we obtain $u_1' = -x \ln x$ and $u_2' = x$. Then $u_1 = \frac{1}{4}x^2 - \frac{1}{2}x^2 \ln x$, $u_2 = \frac{1}{2}x^2$, and

$$y = c_1 + c_2 \ln x + \frac{1}{4}x^2 - \frac{1}{2}x^2 \ln x + \frac{1}{2}x^2 \ln x = c_1 + c_2 \ln x + \frac{1}{4}x^2.$$

The initial conditions imply $c_1 + \frac{1}{4} = 1$ and $c_2 + \frac{1}{2} = -\frac{1}{2}$. Thus, $c_1 = \frac{3}{4}$, $c_2 = -1$, and $y = \frac{3}{4} - \ln x + \frac{1}{4}x^2$.

30. Substituting into the differential equation we obtain

$$\frac{d^2y}{dt^2} - 10\frac{dy}{dt} + 25y = 0.$$

The auxiliary equation is $m^2 - 10m + 25 = (m-5)^2 = 0$ so that

$$y = c_1 e^{5t} + c_2 t e^{5t} = c_1 x^5 + c_2 x^5 \ln x.$$

33. Substituting into the differential equation we obtain

$$\frac{d^2y}{dt^2} - 4\frac{dy}{dt} + 13y = 4 + 3e^t.$$

The auxiliary equation is $m^2 - 4m + 13 = 0$ so that $y_c = e^{2t}(c_1 \cos 3t + c_2 \sin 3t)$. Using undetermined coefficients we try $y_p = A + Be^t$. This leads to $13A + 10Be^t = 4 + 3e^t$, so that $A = 4/13$, $B = 3/10$, and

$$y = e^{2t}(c_1 \cos 3t + c_2 \sin 3t) + \frac{4}{13} + \frac{3}{10}e^t$$

$$= x^2 [c_1 \cos(3 \ln x) + c_2 \sin(3 \ln x)] + \frac{4}{13} + \frac{3}{10}x.$$

36. The differential equation and initial conditions become

$$t^2 \frac{d^2y}{dt^2} - 4t\frac{dy}{dt} + 6y = 0; \quad y(t)\Big|_{t=2} = 8, \quad y'(t)\Big|_{t=2} = 0.$$

The auxiliary equation is $m^2 - 5m + 6 = (m-2)(m-3) = 0$, so that

$$y = c_1 t^2 + c_2 t^3 \quad \text{and} \quad y' = 2c_1 t + 3c_2 t^2.$$

The initial conditions imply

$$4c_1 + 8c_2 = 8$$

$$4c_1 + 12c_2 = 0$$

from which we find $c_1 = 6$ and $c_2 = -2$. Thus

$$y = 6t^2 - 2t^3 = 6x^2 + 2x^3, \quad x < 0.$$

Exercises 4.8

3. From $Dx = -y + t$ and $Dy = x - t$ we obtain $y = t - Dx$, $Dy = 1 - D^2x$, and $(D^2 + 1)x = 1 + t$. Then

$$x = c_1 \cos t + c_2 \sin t + 1 + t$$

and

$$y = c_1 \sin t - c_2 \cos t + t - 1.$$

6. From $(D + 1)x + (D - 1)y = 2$ and $3x + (D + 2)y = -1$ we obtain $x = -\frac{1}{3} - \frac{1}{3}(D + 2)y$, $Dx = -\frac{1}{3}(D^2 + 2D)y$, and $(D^2 + 5)y = -7$. Then

$$y = c_1 \cos \sqrt{5}\,t + c_2 \sin \sqrt{5}\,t - \frac{7}{5}$$

and

$$x = \left(-\frac{2}{3}c_1 - \frac{\sqrt{5}}{3}c_2\right) \cos \sqrt{5}\,t + \left(\frac{\sqrt{5}}{3}c_1 - \frac{2}{3}c_2\right) \sin \sqrt{5}\,t + \frac{3}{5}.$$

9. From $Dx + D^2y = e^{3t}$ and $(D + 1)x + (D - 1)y = 4e^{3t}$ we obtain $D(D^2 + 1)x = 34e^{3t}$ and $D(D^2 + 1)y = -8e^{3t}$. Then

$$y = c_1 + c_2 \sin t + c_3 \cos t - \frac{4}{15}e^{3t}$$

and

$$x = c_4 + c_5 \sin t + c_6 \cos t + \frac{17}{15}e^{3t}.$$

Substituting into $(D + 1)x + (D - 1)y = 4e^{3t}$ gives

$$(c_4 - c_1) + (c_5 - c_6 - c_3 - c_2) \sin t + (c_6 + c_5 + c_2 - c_3) \cos t = 0$$

so that $c_4 = c_1$, $c_5 = c_3$, $c_6 = -c_2$, and

$$x = c_1 - c_2 \cos t + c_3 \sin t + \frac{17}{15}e^{3t}.$$

12. From $(2D^2 - D - 1)x - (2D + 1)y = 1$ and $(D - 1)x + Dy = -1$ we obtain $(2D + 1)(D - 1)(D + 1)x = -1$ and $(2D + 1)(D + 1)y = -2$. Then

$$x = c_1 e^{-t/2} + c_2 e^{-t} + c_3 e^t + 1$$

and

$$y = c_4 e^{-t/2} + c_5 e^{-t} - 2.$$

Substituting into $(D - 1)x + Dy = -1$ gives

$$\left(-\frac{3}{2}c_1 - \frac{1}{2}c_4\right) e^{-t/2} + (-2c_2 - c_5)e^{-t} = 0$$

so that $c_4 = -3c_1$, $c_5 = -2c_2$, and

$$y = -3c_1 e^{-t/2} - 2c_2 e^{-t} - 2.$$

15. Multiplying the first equation by $D + 1$ and the second equation by $D^2 + 1$ and subtracting we obtain $(D^4 - D^2)x = 1$. Then

$$x = c_1 + c_2 t + c_3 e^t + c_4 e^{-t} - \frac{1}{2}t^2.$$

Multiplying the first equation by $D + 1$ and subtracting we obtain $D^2(D + 1)y = 1$. Then

$$y = c_5 + c_6 t + c_7 e^{-t} - \frac{1}{2}t^2.$$

Substituting into $(D - 1)x + (D^2 + 1)y = 1$ gives

$$(-c_1 + c_2 + c_5 - 1) + (-2c_4 + 2c_7)e^{-t} + (-1 - c_2 + c_6)t = 1$$

so that $c_5 = c_1 - c_2 + 2$, $c_6 = c_2 + 1$, and $c_7 = c_4$. The solution of the system is

$$x = c_1 + c_2 t + c_3 e^t + c_4 e^{-t} - \frac{1}{2}t^2$$

$$y = (c_1 - c_2 + 2) + (c_2 + 1)t + c_4 e^{-t} - \frac{1}{2}t^2.$$

18. From $Dx + z = e^t$, $(D - 1)x + Dy + Dz = 0$, and $x + 2y + Dz = e^t$ we obtain $z = -Dx + e^t$, $Dz = -D^2 x + e^t$, and the system $(-D^2 + D - 1)x + Dy = -e^t$ and $(-D^2 + 1)x + 2y = 0$. Then $y = \frac{1}{2}(D^2 - 1)x$, $Dy = \frac{1}{2}D(D^2 - 1)x$, and $(D - 2)(D^2 + 1)x = -2e^t$ so that

$$x = c_1 e^{2t} + c_2 \cos t + c_3 \sin t + e^t,$$

$$y = \frac{3}{2}c_1 e^{2t} - c_2 \cos t - c_3 \sin t,$$

and

$$z = -2c_1 e^{2t} - c_3 \cos t + c_2 \sin t.$$

21. From $(D + 5)x + y = 0$ and $4x - (D + 1)y = 0$ we obtain $y = -(D + 5)x$ so that $Dy = -(D^2 + 5D)x$. Then $4x + (D^2 + 5D)x + (D + 5)x = 0$ and $(D + 3)^2 x = 0$. Thus

$$x = c_1 e^{-3t} + c_2 t e^{-3t}$$

and

$$y = -(2c_1 + c_2)e^{-3t} - 2c_2 t e^{-3t}.$$

Using $x(1) = 0$ and $y(1) = 1$ we obtain

$$c_1 e^{-3} + c_2 e^{-3} = 0$$

$$-(2c_1 + c_2)e^{-3} - 2c_2 e^{-3} = 1$$

45

or

$$c_1 + c_2 = 0$$

$$2c_1 + 3c_2 = -e^3.$$

Thus $c_1 = e^3$ and $c_2 = -e^3$. The solution of the initial value problem is

$$x = e^{-3t+3} - te^{-3t+3}$$

$$y = -e^{-3t+3} + 2te^{-3t+3}.$$

24. From Newton's second law in the x-direction we have

$$m\frac{d^2x}{dt^2} = -k\cos\theta = -k\frac{1}{v}\frac{dx}{dt} = -|c|\frac{dx}{dt}.$$

In the y-direction we have

$$m\frac{d^2y}{dt^2} = -mg - k\sin\theta = -mg - k\frac{1}{v}\frac{dy}{dt} = -mg - |c|\frac{dy}{dt}.$$

From $mD^2x + |c|Dx = 0$ we have $D(mD + |c|)x = 0$ so that $(mD + |c|)x = c_1$. This is a first-order linear equation. An integrating factor is $e^{\int |c|dt/m}e^{|c|t/m}$ so that

$$\frac{d}{dt}[e^{|c|t/m}x] = c_1e^{|c|t/m}$$

and $e^{|c|t}x = (c_1m/|c|)e^{|c|t/m}+c_2$. The general solution of this equation is $x(t) = c_3+c_2e^{|c|t/m}$. From $(mD^2 + |c|D)y = -mg$ we have $D(mD + |c|)y = -mg$ so that $(mD + |c|)y = -mgt + c_1$. This is a first-order linear equation with integrating factor $e^{|c|t/m}$. Thus

$$\frac{d}{dt}[e^{|c|t/m}y] = (-mgt + c_1)e^{|c|t/m}$$

$$e^{|c|t/m}y = -\frac{m^2g}{|c|}te^{|c|t/m} + \frac{m^3g}{c^2}e^{|c|t/m} + \frac{c_1m}{|c|}e^{|c|t/m} + c_2$$

and

$$y(t) = -\frac{m^2g}{|c|}t + \frac{m^3g}{c^2} + c_3 + c_2e^{-|c|t/m}.$$

—————— **Exercises 4.9** ——————

3. Let $u = y'$ so that $u' = y''$. The equation becomes $u' = -u - 1$ which is separable. Thus

$$\frac{du}{u^2 + 1} = -dx \implies \tan^{-1}u = -x + c_1 \implies y' = \tan(c_1 - x) \implies y = \ln|\cos(c_1 - x)| + c_2.$$

6. Let $u = y'$ so that $y'' = u\dfrac{du}{dy}$. The equation becomes $(y+1)u\dfrac{du}{dy} = u^2$. Separating variables we obtain

$$\frac{du}{u} = \frac{dy}{y+1} \implies \ln|u| = \ln|y+1| + \ln c_1 \implies u = c_1(y+1)$$

$$\implies \frac{dy}{dx} = c_1(y+1) \implies \frac{dy}{y+1} = c_1\,dx$$

$$\implies \ln|y+1| = c_1 x + c_2 \implies y+1 = c_3 e^{c_1 x}.$$

9. (a)

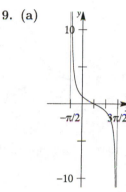

(b) Let $u = y'$ so that $y'' = u\dfrac{du}{dy}$. The equation becomes $u\dfrac{du}{dy} + yu = 0$. Separating variables we obtain

$$du = -y\,dy \implies u = -\frac{1}{2}y^2 + c_1 \implies y' = -\frac{1}{2}y^2 + c_1.$$

When $x = 0$, $y = 1$ and $y' = -1$ so $-1 = -\frac{1}{2} + c_1$ and $c_1 = -\frac{1}{2}$. Then

$$\frac{dy}{dx} = -\frac{1}{2}y^2 - \frac{1}{2} \implies \frac{dy}{y^2+1} = -\frac{1}{2}\,dx \implies \tan^{-1} y = -\frac{1}{2}x + c_2$$

$$\implies y = \tan\left(-\frac{1}{2}x + c_2\right).$$

When $x = 0$, $y = 1$ so $1 = \tan c_2$ and $c_2 = \pi/4$. The solution of the initial-value problem is

$$y = \tan\left(\frac{\pi}{4} - \frac{1}{2}x\right).$$

The graph is shown in part (a).

(c) The interval of defintion is $-\pi/2 < \pi/4 - x/2 < \pi/2$ or $-\pi/2 < x < 3\pi/2$.

12. Let $u = y'$ so that $u' = y''$. The equation becomes $u' - \dfrac{1}{x}u = u^2$, which is Bernoulli. Using the substitution $w = u^{-1}$ we obtain $\dfrac{dw}{dx} + \dfrac{1}{x}w = -1$. An integrating factor is x, so

$$\frac{d}{dx}[xw] = -x \implies w = -\frac{1}{2}x + \frac{1}{x}c \implies \frac{1}{u} = \frac{c_1 - x^2}{2x} \implies u = \frac{2x}{c_1 - x^2} \implies y = -\ln\left|c_1 - x^2\right| + c_2.$$

47

15. We look for a solution of the form

$$y(x) = y(0) + y'(0) + \frac{1}{2}y''(0) + \frac{1}{3!}y'''(0) + \frac{1}{4!}y^{(4)}(x) + \frac{1}{5!}y^{(5)}(x).$$

From $y''(x) = x^2 + y^2 - 2y'$ we compute

$$y'''(x) = 2x + 2yy' - 2y''$$

$$y^{(4)}(x) = 2 + 2(y')^2 + 2yy'' - 2y'''$$

$$y^{(5)}(x) = 6y'y'' + 2yy''' - 2y^{(4)}.$$

Using $y(0) = 1$ and $y'(0) = 1$ we find

$$y''(0) = -1, \quad y'''(0) = 4, \quad y^{(4)}(0) = -6, \quad y^{(5)}(0) = 14.$$

An approximate solution is

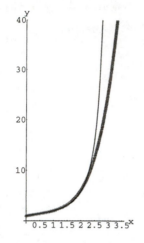

$$y(x) = 1 + x - \frac{1}{2}x^2 + \frac{2}{3}x^3 - \frac{1}{4}x^4 + \frac{7}{60}x^5.$$

(The thinner curve is obtained using a numerical solver, while the thicker curve is the graph of the Taylor polynomial.)

18. Let $u = \dfrac{dx}{dt}$ so that $\dfrac{d^2x}{dt^2} = u\dfrac{du}{dx}$. The equation becomes $u\dfrac{du}{dx} = \dfrac{-k^2}{x^2}$. Separating variables we obtain

$$u\,du = -\frac{k^2}{x^2}\,dx \implies \frac{1}{2}u^2 = \frac{k^2}{x} + c \implies \frac{1}{2}v^2 = \frac{k^2}{x} + c.$$

When $t = 0$, $x = x_0$ and $v = 0$ so $0 = \dfrac{k^2}{x_0} + c$ and $c = -\dfrac{k^2}{x_0}$. Then

$$\frac{1}{2}v^2 = k^2\left(\frac{1}{x} - \frac{1}{x_0}\right) \quad \text{and} \quad \frac{dx}{dt} = -k\sqrt{2}\sqrt{\frac{x_0 - x}{xx_0}}.$$

Separating variables we have

$$-\sqrt{\frac{xx_0}{x_0 - x}}\,dx = k\sqrt{2}\,dt \implies t = -\frac{1}{k}\sqrt{\frac{x_0}{2}}\int\sqrt{\frac{x}{x_0 - x}}\,dx.$$

Using *Mathematica* to integrate we obtain

$$t = -\frac{1}{k}\sqrt{\frac{x_0}{2}}\left[-\sqrt{x(x_0 - x)} - \frac{x_0}{2}\tan^{-1}\frac{(x_0 - 2x)}{2x}\sqrt{\frac{x}{x_0 - x}}\right]$$

$$= \frac{1}{k}\sqrt{\frac{x_0}{2}}\left[\sqrt{x(x_0 - x)} + \frac{x_0}{2}\tan^{-1}\frac{x_0 - 2x}{2\sqrt{x(x_0 - x)}}\right].$$

48

3. True

6. (a) Since $f_2(x) = 2\ln x = 2f_1(x)$, the functions are linearly dependent.

(b) Since x^{n+1} is not a constant multiple of x^n, the functions are linearly independent.

(c) Since $x + 1$ is not a constant multiple of x, the functions are linearly independent.

(d) Since $f_1(x) = \cos x \cos(\pi/2) - \sin x \sin(\pi/2) = -\sin x = -f_2(x)$, the functions are linearly dependent.

(e) Since $f_1(x) = 0 \cdot f_2(x)$, the functions are linearly dependent.

(f) Since $2x$ is not a constant multiple of 2, the functions are linearly independent.

(g) Since $3(x^2) + 2(1 - x^2) - (2 + x^2) = 0$, the functions are linearly dependent.

(h) Since $xe^{x+1} + 0(4x - 5)e^x - exe^x = 0$, the functions are linearly dependent.

9. From $m^2 - 2m - 2 = 0$ we obtain $m = 1 \pm \sqrt{3}$ so that

$$y = c_1 e^{(1+\sqrt{3})x} + c_2 e^{(1-\sqrt{3})x}.$$

12. From $2m^3 + 9m^2 + 12m + 5 = 0$ we obtain $m = -1$, $m = -1$, and $m = -5/2$ so that

$$y = c_1 e^{-5x/2} + c_2 e^{-x} + c_3 xe^{-x}.$$

15. Applying D^4 to the differential equation we obtain $D^4(D^2 - 3D + 5) = 0$. Then

$$y = \underbrace{e^{3x/2}\left(c_1 \cos\frac{\sqrt{11}}{2}x + c_2 \sin\frac{\sqrt{11}}{2}x\right)}_{y_c} + c_3 + c_4 x + c_5 x^2 + c_6 x^3$$

and $y_p = A + Bx + Cx^2 + Dx^3$. Substituting y_p into the differential equation yields

$$(5A - 3B + 2C) + (5B - 6C + 6D)x + (5C - 9D)x^2 + 5Dx^3 = -2x + 4x^3.$$

Equating coefficients gives $A = -222/625$, $B = 46/125$, $C = 36/25$, and $D = 4/5$. The general solution is

$$y = e^{3x/2}\left(c_1 \cos\frac{\sqrt{11}}{2}x + c_2 \sin\frac{\sqrt{11}}{2}x\right) - \frac{222}{625} + \frac{46}{125}x + \frac{36}{25}x^2 + \frac{4}{5}x^3.$$

18. Applying D to the differential equation we obtain $D(D^3 - D^2) = D^3(D - 1) = 0$. Then

$$y = \underbrace{c_1 + c_2 x + c_3 e^x}_{y_c} + c_4 x^2$$

and $y_p = Ax^2$. Substituting y_p into the differential equation yields $-2A = 6$. Equating coefficients gives $A = -3$. The general solution is

$$y = c_1 + c_2 x + c_3 e^x - 3x^2.$$

49

21. The auxiliary equation is $6m^2 - m - 1 = 0$ so that

$$y = c_1 x^{1/2} + c_2 x^{-1/3}.$$

24. The auxiliary equation is $m^2 - 2m + 1 = (m-1)^2 = 0$ and a particular solution is $y_p = \frac{1}{4}x^3$ so that

$$y = c_1 x + c_2 x \ln x + \frac{1}{4}x^3.$$

27. (a) The auxiliary equation is $m^4 - 2m^2 + 1 = (m^2 - 1)^2 = 0$, so the general solution of the differential equation is

$$y = c_1 \sinh x + c_2 \cosh x + c_3 x \sinh x + c_4 x \cosh x.$$

(b) Since both $\sinh x$ and $x \sinh x$ are solutions of the associated homogeneous differential equation, a particular solution of $y^{(4)} - 2y'' + y = \sinh x$ has the form $y_p = Ax^2 \sinh x + Bx^2 \cosh x$.

30. The auxiliary equation is $m^2 + 2m + 1 = (m+1)^2 = 0$, so that $y = c_1 e^{-x} + c_2 x e^{-x}$. Setting $y(-1) = 0$ and $y'(0) = 0$ we get $c_1 e - c_2 e = 0$ and $-c_1 + c_2 = 0$. Thus $c_1 = c_2$ and $y = ce^{-x} + cxe^{-x}$ is a solution of the boundary-value problem for any real number c.

33. Let $u = y'$ so that $u' = y''$. The equation becomes $u\dfrac{du}{dx} = 4x$. Separating variables we obtain

$$u\,du = 4x\,dx \implies \frac{1}{2}u^2 = 2x^2 + c_1 \implies u^2 = 4x^2 + c_2.$$

When $x = 1$, $y' = u = 2$, so $4 = 4 + c_2$ and $c_2 = 0$. Then

$$u^2 = 4x^2 \implies \frac{dy}{dx} = 2x \quad \text{or} \quad \frac{dy}{dx} = -2x$$

$$\implies y = x^2 + c_3 \quad \text{or} \quad y = -x^2 + c_4.$$

When $x = 1$, $y = 5$, so $5 = 1 + c_3$ and $5 = -1 + c_4$. Thus $c_3 = 4$ and $c_4 = 6$. We have $y = x^2 + 4$ and $y = -x^2 + 6$. Note however that when $y = -x^2 + 6$, $y' = -2x$ and $y'(1) = -2 \neq 2$. Thus, the solution of the initial-value problem is $y = x^2 + 4$.

36. Consider $xy'' + y' = 0$ and look for a solution of the form $y = x^m$. Substituting into the differential equation we have

$$xy'' + y' = m(m-1)x^{m-1} + mx^{m-1} = m^2 x.$$

Thus, the general solution of $xy'' + y' = 0$ is $y_c = c_1 + c_2 \ln x$. To find a particular solution of $xy'' + y' = -\sqrt{x}$ we use variation of parameters.

The Wronskian is

$$W = \begin{vmatrix} 1 & \ln x \\ 0 & 1/x \end{vmatrix} = \frac{1}{x}.$$

Identifying $f(x) = -x^{-1/2}$ we obtain

$$u_1' = \frac{x^{-1/2}\ln x}{1/x} = \sqrt{x}\ln x \quad \text{and} \quad u_2' = \frac{-x^{-1/2}}{1/x} = -\sqrt{x},$$

so that

$$u_1 = x^{3/2}\left(\frac{2}{3}\ln x - \frac{4}{9}\right) \quad \text{and} \quad u_2 = -\frac{2}{3}x^{3/2}.$$

Then $y_p = x^{3/2}(\frac{2}{3}\ln x - \frac{4}{9}) - \frac{2}{3}x^{3/2}\ln x = -\frac{4}{9}x^{3/2}$ and the general solution of the differential equation is $y = c_1 + c_2\ln x - \frac{4}{9}x^{3/2}$. The initial conditions are $y(1) = 0$ and $y'(1) = 0$. These imply that $c_1 = \frac{4}{9}$ and $c_2 = \frac{2}{3}$. The solution of the initial-value problem is $y = \frac{4}{9} + \frac{2}{3}\ln x - \frac{4}{9}x^{3/2}$.

39. From $(D-2)x - y = -e^t$ and $-3x + (D-4)y = -7e^t$ we obtain $(D-1)(D-5)x = -4e^t$ so that

$$x = c_1 e^t + c_2 e^{5t} + te^t.$$

Then

$$y = (D-2)x + e^t = -c_1 e^t + 3c_2 e^{5t} - te^t + 2e^t.$$

5 Modeling with Higher-Order Differential Equations

Exercises 5.1

3. From $\frac{3}{4}x'' + 72x = 0$, $x(0) = -1/4$, and $x'(0) = 0$ we obtain $x = -\frac{1}{4}\cos 4\sqrt{6}\,t$.

6. From $50x'' + 200x = 0$, $x(0) = 0$, and $x'(0) = -10$ we obtain $x = -5\sin 2t$ and $x' = -10\cos 2t$.

9. From $\frac{1}{4}x'' + x = 0$, $x(0) = 1/2$, and $x'(0) = 3/2$ we obtain

$$x = \frac{1}{2}\cos 2t + \frac{3}{4}\sin 2t = \frac{\sqrt{13}}{4}\sin(2t + 0.588).$$

12. From $x'' + 9x = 0$, $x(0) = -1$, and $x'(0) = -\sqrt{3}$ we obtain

$$x = -\cos 3t - \frac{\sqrt{3}}{3}\sin 3t = \frac{2}{\sqrt{3}}\sin\left(37 + \frac{4\pi}{3}\right)$$

and $x' = 2\sqrt{3}\cos(3t + 4\pi/3)$. If $x' = 3$ then $t = -7\pi/18 + 2n\pi/3$ and $t = -\pi/2 + 2n\pi/3$ for $n = 1, 2, 3, \ldots$.

15. For large values of t the differential equation is approximated by $x'' = 0$. The solution of this equation is the linear function $x = c_1 t + c_2$. Thus, for large time, the restoring force will have decayed to the point where the spring is incapable of returning the mass, and the spring will simply keep on stretching.

18. (a) below   (b)   from rest

21. From $\frac{1}{8}x'' + x' + 2x = 0$, $x(0) = -1$, and $x'(0) = 8$ we obtain $x = 4te^{-4t} - e^{-4t}$ and $x' = 8e^{-4t} - 16te^{-4t}$. If $x = 0$ then $t = 1/4$ second. If $x' = 0$ then $t = 1/2$ second and the extreme displacement is $x = e^{-2}$ feet.

24. (a) $x = \frac{1}{3}e^{-8t}\left(4e^{6t} - 1\right)$ is never zero; the extreme displacement is $x(0) = 1$ meter.

    (b) $x = \frac{1}{3}e^{-8t}\left(5 - 2e^{6t}\right) = 0$ when $t = \frac{1}{6}\ln\frac{5}{2} \approx 0.153$ second; if $x' = \frac{4}{3}e^{-8t}\left(e^{6t} - 10\right) = 0$ then $t = \frac{1}{6}\ln 10 \approx 0.384$ second and the extreme displacement is $x = -0.232$ meter.

27. From $\frac{5}{16}x'' + \beta x' + 5x = 0$ we find that the roots of the auxiliary equation are $m = -\frac{8}{5}\beta \pm \frac{4}{5}\sqrt{4\beta^2 - 25}$.

    (a) If $4\beta^2 - 25 > 0$ then $\beta > 5/2$.

    (b) If $4\beta^2 - 25 = 0$ then $\beta = 5/2$.

    (c) If $4\beta^2 - 25 < 0$ then $0 < \beta < 5/2$.

30. (a) If $x'' + 2x' + 5x = 12\cos 2t + 3\sin 2t$, $x(0) = -1$, and $x'(0) = 5$ then $x_c = e^{-t}(c_1 \cos 2t + c_2 \sin 2t)$ and $x_p = 3\sin 2t$ so that the equation of motion is

$$x = e^{-t}\cos 2t + 3\sin 2t.$$

(b)   **(c)**

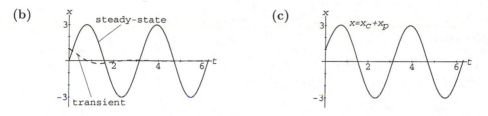

33. From $2x'' + 32x = 68e^{-2t}\cos 4t$, $x(0) = 0$, and $x'(0) = 0$ we obtain $x_c = c_1\cos 4t + c_2\sin 4t$ and $x_p = \frac{1}{2}e^{-2t}\cos 4t - 2e^{-2t}\sin 4t$ so that

$$x = -\frac{1}{2}\cos 4t + \frac{9}{4}\sin 4t + \frac{1}{2}e^{-2t}\cos 4t - 2e^{-2t}\sin 4t.$$

36. (a) From $100x'' + 1600x = 1600\sin 8t$, $x(0) = 0$, and $x'(0) = 0$ we obtain $x_c = c_1\cos 4t + c_2\sin 4t$ and $x_p = -\frac{1}{3}\sin 8t$ so that

$$x = \frac{2}{3}\sin 4t - \frac{1}{3}\sin 8t.$$

(b) If $x = \frac{1}{3}\sin 4t(2 - 2\cos 4t) = 0$ then $t = n\pi/4$ for $n = 0, 1, 2, \ldots$.

(c) If $x' = \frac{8}{3}\cos 4t - \frac{8}{3}\cos 8t = \frac{8}{3}(1-\cos 4t)(1+2\cos 4t) = 0$ then $t = \pi/3 + n\pi/2$ and $t = \pi/6 + n\pi/2$ for $n = 0, 1, 2, \ldots$ at the extreme values. *Note*: There are many other values of t for which $x' = 0$.

(d) $x(\pi/6 + n\pi/2) = \sqrt{3}/2$ cm. and $x(\pi/3 + n\pi/2) = -\sqrt{3}/2$ cm.

(e)

39. (a) From $x'' + \omega^2 x = F_0\cos\gamma t$, $x(0) = 0$, and $x'(0) = 0$ we obtain $x_c = c_1\cos\omega t + c_2\sin\omega t$ and $x_p = (F_0\cos\gamma t)/(\omega^2 - \gamma^2)$ so that

$$x = -\frac{F_0}{\omega^2 - \gamma^2}\cos\omega t + \frac{F_0}{\omega^2 - \gamma^2}\cos\gamma t.$$

(b) $\displaystyle\lim_{\gamma\to\omega} \frac{F_0}{\omega^2 - \gamma^2}(\cos\gamma t - \cos\omega t) = \lim_{\gamma\to\omega} \frac{-F_0 t\sin\gamma t}{-2\gamma} = \frac{F_0}{2\omega}t\sin\omega t.$

53

45. Solving $\frac{1}{20}q'' + 2q' + 100q = 0$ we obtain $q(t) = e^{-20t}(c_1 \cos 40t + c_2 \sin 40t)$. The initial conditions $q(0) = 5$ and $q'(0) = 0$ imply $c_1 = 5$ and $c_2 = 5/2$. Thus

$$q(t) = e^{-20t}\left(5 \cos 40t + \frac{5}{2} \sin 40t\right) \approx \sqrt{25 + 25/4}\, e^{-20t} \sin(40t + 1.1071)$$

and $q(0.01) \approx 4.5676$ coulombs. The charge is zero for the first time when $40t + 0.4636 = \pi$ or $t \approx 0.0509$ second.

48. Solving $q'' + 100q' + 2500q = 30$ we obtain $q(t) = c_1 e^{-50t} + c_2 t e^{-50t} + 0.012$. The initial conditions $q(0) = 0$ and $q'(0) = 2$ imply $c_1 = -0.012$ and $c_2 = 1.4$. Thus

$$q(t) = -0.012e^{-50t} + 1.4te^{-50t} + 0.012 \quad \text{and} \quad i(t) = 2e^{-50t} - 70te^{-50t}.$$

Solving $i(t) = 0$ we see that the maximum charge occurs when $t = 1/35$ and $q(1/35) \approx 0.01871$.

51. The differential equation is $\frac{1}{2}q'' + 20q' + 1000q = 100 \sin t$. To use Example 10 in the text we identify $E_0 = 100$ and $\gamma = 60$. Then

$$X = L\gamma - \frac{1}{c\gamma} = \frac{1}{2}(60) - \frac{1}{0.001(60)} \approx 13.3333,$$

$$Z = \sqrt{X^2 + R^2} = \sqrt{X^2 + 400} \approx 24.0370,$$

and

$$\frac{E_0}{Z} = \frac{100}{Z} \approx 4.1603.$$

From Problem 50, then

$$i_p(t) \approx 4.1603(60t + \phi)$$

where $\sin \phi = -X/Z$ and $\cos \phi = R/Z$. Thus $\tan \phi = -X/R \approx -0.6667$ and ϕ is a fourth quadrant angle. Now $\phi \approx -0.5880$ and

$$i_p(t) \approx 4.1603(60t - 0.5880).$$

54. By Problem 50 the amplitude of the steady-state current is E_0/Z, where $Z = \sqrt{X^2 + R^2}$ and $X = L\gamma - 1/C\gamma$. Since E_0 is constant the amplitude will be a maximum when Z is a minimum. Since R is constant, Z will be a minimum when $X = 0$. Solving $L\gamma - 1/C\gamma = 0$ for γ we obtain $\gamma = 1/\sqrt{LC}$. The maximum amplitude will be E_0/R.

57. In an L-C series circuit there is no resistor, so the differential equation is

$$L\frac{d^2q}{dt^2} + \frac{1}{C}q = E(t).$$

Then $q(t) = c_1 \cos\left(t/\sqrt{LC}\right) + c_2 \sin\left(t/\sqrt{LC}\right) + q_p(t)$ where $q_p(t) = A \sin \gamma t + B \cos \gamma t$. Substituting $q_p(t)$ into the differential equation we find

$$\left(\frac{1}{C} - L\gamma^2\right) A \sin \gamma t + \left(\frac{1}{C} - L\gamma^2\right) B \cos \gamma t = E_0 \cos \gamma t.$$

Equating coefficients we obtain $A = 0$ and $B = E_0 C/(1 - LC\gamma^2)$. Thus, the charge is

$$q(t) = c_1 \cos \frac{1}{\sqrt{LC}} t + c_2 \sin \frac{1}{\sqrt{LC}} t + \frac{E_0 C}{1 - LC\gamma^2} \cos \gamma t.$$

The initial conditions $q(0) = q_0$ and $q'(0) = i_0$ imply $c_1 = q_0 - E_0 C/(1 - LC\gamma^2)$ and $c_2 = i_0 \sqrt{LC}$. The current is

$$i(t) = -\frac{c_1}{\sqrt{LC}} \sin \frac{1}{\sqrt{LC}} t + \frac{c_2}{\sqrt{LC}} \cos \frac{1}{\sqrt{LC}} t - \frac{E_0 C\gamma}{1 - LC\gamma^2} \sin \gamma t$$

$$= i_0 \cos \frac{1}{\sqrt{LC}} t - \frac{1}{\sqrt{LC}} \left(q_0 - \frac{E_0 C}{1 - LC\gamma^2} \right) \sin \frac{1}{\sqrt{LC}} t - \frac{E_0 C\gamma}{1 - LC\gamma^2} \sin \gamma t.$$

Exercises 5.2

3. (a) The general solution is

$$y(x) = c_1 + c_2 x + c_3 x^2 + c_4 x^3 + \frac{w_0}{24EI} x^4.$$

The boundary conditions are $y(0) = 0$, $y'(0) = 0$, $y(L) = 0$, $y''(L) = 0$. The first two conditions give $c_1 = 0$ and $c_2 = 0$. The conditions at $x = L$ give the system

$$c_3 L^2 + c_4 L^3 + \frac{w_0}{24EI} L^4 = 0$$

$$2c_3 + 6c_4 L + \frac{w_0}{2EI} L^2 = 0.$$

Solving, we obtain $c_3 = w_0 L^2/16EI$ and $c_4 = -5w_0 L/48EI$. The deflection is

$$y(x) = \frac{w_0}{48EI} (3L^2 x^2 - 5Lx^3 + 2x^4).$$

(b)

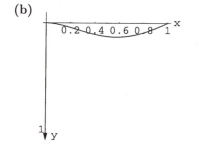

6. (a) The deflection of the beam in Problem 1 is

$$y(x) = \frac{w_0}{24EI} (6L^2 x^2 - 4Lx^3 + x^4).$$

Since $y(x)$ is a differentiable function on $[0, L]$ we can find the maximum deflection by comparing the deflections at $x = 0$, $x = L$, and at any critical points. Setting

$$y' = \frac{w_0 x^2}{24EI}(x^2 - 4Lx + 6L^2) = 0$$

and noting that $x^2 - 4Lx + 6L^2$ has no real roots we see that $x = 0$ is the only critical point. Since $y(0) = 0$, the maximum deflection is $y_{max} = y(L) = w_0 L^4/8EI$.

(b) Replacing both L and x by $L/2$ in $y(x)$ we obtain $w_0 L^4/128EI$, which is $1/16$ of the maximum deflection when the length of the beam is L.

(c) The deflection of the beam in Problem 2 is

$$y(x) = \frac{w_0}{24EI}(L^3 x - 2Lx^3 + x^4).$$

Since $y(x)$ is a differentiable function on $[0, L]$ we can find the maximum deflection by comparing the deflections at $x = 0$, $x = L$, and at any critical points. Setting

$$y' = \frac{w_0}{24EI}(4x^3 - 6Lx^2 + L^3) = 0,$$

we see that $x = L/2$ is the only critical point in $[0, L]$. Since $y(0) = y(L) = 0$, the maximum deflection is $y_{max} = y(L/2) = 5w_0 L^4/384EI$.

(d) The maximum deflection in Example 1 is $y(L/2) = (w_0/24EI)L^4/16 = w_0 L^4/384EI$, which is $1/5$ of the maximum displacement of the beam in Problem 2.

9. For $\lambda \leq 0$ the only solution of the boundary-value problem is $y = 0$. For $\lambda > 0$ we have

$$y = c_1 \cos \sqrt{\lambda}\, x + c_2 \sin \sqrt{\lambda}\, x.$$

Now $y(0) = 0$ implies $c_1 = 0$, so

$$y(\pi) = c_2 \sin \sqrt{\lambda}\, \pi = 0$$

gives

$$\sqrt{\lambda}\, \pi = n\pi \quad \text{or} \quad \lambda = n^2, \ n = 1, 2, 3, \ldots .$$

The eigenvalues n^2 correspond to the eigenfunctions $\sin nx$ for $n = 1, 2, 3, \ldots$.

12. For $\lambda \leq 0$ the only solution of the boundary-value problem is $y = 0$. For $\lambda > 0$ we have

$$y = c_1 \cos \sqrt{\lambda}\, x + c_2 \sin \sqrt{\lambda}\, x.$$

Now $y(0) = 0$ implies $c_1 = 0$, so

$$y'\left(\frac{\pi}{2}\right) = c_2 \sqrt{\lambda} \cos \sqrt{\lambda}\, \frac{\pi}{2} = 0$$

gives

$$\sqrt{\lambda}\, \frac{\pi}{2} = \frac{(2n-1)\pi}{2} \quad \text{or} \quad \lambda = (2n-1)^2, \ n = 1, 2, 3, \ldots .$$

The eigenvalues $(2n-1)^2$ correspond to the eigenfunctions $\sin(2n-1)x$.

15. The auxiliary equation has solutions

$$m = \frac{1}{2}\left(-2 \pm \sqrt{4-4(\lambda+1)}\right) = -1 \pm \sqrt{-\lambda}.$$

For $\lambda < 0$ we have

$$y = e^{-x}\left(c_1 \cosh\sqrt{-\lambda}\,x + c_2 \sinh\sqrt{-\lambda}\,x\right).$$

The boundary conditions imply

$$y(0) = c_1 = 0$$

$$y(5) = c_2 e^{-5}\sinh 5\sqrt{-\lambda} = 0$$

so $c_1 = c_2 = 0$ and the only solution of the boundary-value problem is $y = 0$.
For $\lambda = 0$ we have

$$y = c_1 e^{-x} + c_2 x e^{-x}$$

and the only solution of the boundary-value problem is $y = 0$.
For $\lambda > 0$ we have

$$y = e^{-x}\left(c_1 \cos\sqrt{\lambda}\,x + c_2 \sin\sqrt{\lambda}\,x\right).$$

Now $y(0) = 0$ implies $c_1 = 0$, so

$$y(5) = c_2 e^{-5}\sin 5\sqrt{\lambda} = 0$$

gives

$$5\sqrt{\lambda} = n\pi \quad\text{or}\quad \lambda = \frac{n^2\pi^2}{25},\; n = 1,2,3,\ldots.$$

The eigenvalues $n^2\pi^2/25$ correspond to the eigenfunctions $e^{-x}\sin\frac{n\pi}{5}x$ for $n = 1, 2, 3, \ldots$.

18. For $\lambda = 0$ the only solution of the boundary-value problem is $y = 0$. For $\lambda \neq 0$ we have

$$y = c_1 \cos\lambda x + c_2 \sin\lambda x.$$

Now $y(0) = 0$ implies $c_1 = 0$, so

$$y'(3\pi) = c_2\lambda\cos 3\pi\lambda = 0$$

gives

$$3\pi\lambda = \frac{(2n-1)\pi}{2} \quad\text{or}\quad \lambda = \frac{2n-1}{6},\; n = 1,2,3,\ldots.$$

The eigenvalues $(2n-1)/6$ correspond to the eigenfunctions $\sin\frac{2n-1}{6}x$ for $n = 1, 2, 3, \ldots$.

21. For $\lambda = 0$ the general solution is $y = c_1 + c_2\ln x$. Now $y' = c_2/x$, so $y'(1) = c_2 = 0$ and $y = c_1$. Since $y'(e^2) = 0$ for any c_1 we see that $y(x) = 1$ is an eigenfunction corresponding to the eigenvalue $\lambda = 0$.
For $\lambda < 0$, $y = c_1 x^{-\sqrt{-\lambda}} + c_2 x^{\sqrt{-\lambda}}$. The initial conditions imply $c_1 = c_2 = 0$, so $y(x) = 0$.

57

For $\lambda > 0$, $y = c_1 \cos(\sqrt{\lambda} \ln x) + c_2 \sin(\sqrt{\lambda} \ln x)$. Now

$$y' = -c_1 \frac{\sqrt{\lambda}}{x} \sin(\sqrt{\lambda} \ln x) + c_2 \frac{\sqrt{\lambda}}{x} \cos(\sqrt{\lambda} \ln x),$$

and $y'(1) = c_2\sqrt{\lambda} = 0$ implies $c_2 = 0$. Finally, $y'(e^2) = -(c_1\sqrt{\lambda}/e^2) \sin(2\sqrt{\lambda}) = 0$ implies $\lambda = n^2\pi^2/4$ for $n = 1, 2, 3, \ldots$. The corresponding eigenfunctions are

$$y = \cos\left(\frac{n\pi}{2} \ln x\right).$$

24. (a) The general solution of the differential equation is

$$y = c_1 \cos\sqrt{\frac{P}{EI}}\, x + c_2 \sin\sqrt{\frac{P}{EI}}\, x + \delta.$$

Since the column is embedded at $x = 0$, the initial conditions are $y(0) = y'(0) = 0$. If $\delta = 0$ this implies that $c_1 = c_2 = 0$ and $y(x) = 0$. That is, there is no deflection.

(b) If $\delta \neq 0$, the initial conditions give, in turn, $c_1 = -\delta$ and $c_2 = 0$. Then

$$y = \delta\left(1 - \cos\sqrt{\frac{P}{EI}}\, x\right).$$

In order to satisfy the condition $y(L) = \delta$ we must have

$$\delta = \delta\left(1 - \cos\sqrt{\frac{P}{EI}}\, L\right) \quad \text{or} \quad \cos\sqrt{\frac{P}{EI}}\, L = 0.$$

This gives $\sqrt{P/EI}\, L = n\pi/2$ for $n = 1, 2, 3, \ldots$. The smallest value of P_n, the Euler load, is then

$$\sqrt{\frac{P_1}{EI}}\, L = \frac{\pi}{2} \quad \text{or} \quad P_1 = \frac{1}{4}\left(\frac{\pi^2 EI}{L^2}\right).$$

27. The auxiliary equation is $m^2 + m = m(m+1) = 0$ so that $u(r) = c_1 r^{-1} + c_2$. The boundary conditions $u(a) = u_0$ and $u(b) = u_1$ yield the system $c_1 a^{-1} + c_2 = u_0$, $c_1 b^{-1} + c_2 = u_1$. Solving gives

$$c_1 = \left(\frac{u_0 - u_1}{b - a}\right) ab \quad \text{and} \quad c_2 = \frac{u_1 b - u_0 a}{b - a}.$$

Thus

$$u(r) = \left(\frac{u_0 - u_1}{b - a}\right)\frac{ab}{r} + \frac{u_1 b - u_0 a}{b - a}.$$

3. The period corresponding to $x(0) = 1$, $x'(0) = 1$ is approximately 5.8. The second initial-value problem does not have a periodic solution.

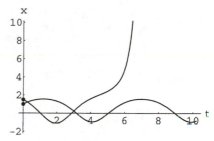

6. From the graphs we see that the interval is approximately $(-0.8, 1.1)$.

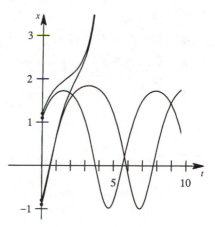

9. This is a damped hard spring, so all solutions should be oscillatory with $x \to 0$ as $t \to \infty$.

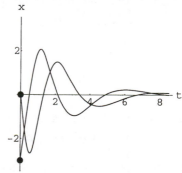

12. (a)

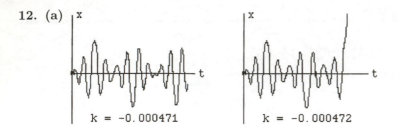

The system appears to be oscillatory for $-0.000471 \leq k_1 < 0$ and nonoscillatory for $k_1 \leq -0.000472$.

(b)

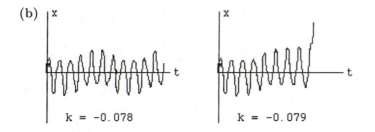

The system appears to be oscillatory for $-0.077 \leq k_1 < 0$ and nonoscillatory for $k_1 \leq 0.078$.

15. (a) Intuitively, one might expect that only half of a 10-pound chain could be lifted by a 5-pound force.

(b) Since $x = 0$ when $t = 0$, and $v = dx/dt = \sqrt{160 - 64x/3}$, we have $v(0) = \sqrt{160} \approx 12.65$ ft/s.

(c) Since x should always be positive, we solve $x(t) = 0$, getting $t = 0$ and $t = \frac{3}{2}\sqrt{5/2} \approx 2.3717$. Since the graph of $x(t)$ is a parabola, the maximum value occurs at $t_m = \frac{3}{4}\sqrt{5/2}$. (This can also be obtained by solving $x'(t) = 0$.) At this time the height of the chain is $x(t_m) \approx 7.5$ ft. This is higher than predicted because

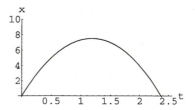

of the momentum generated by the force. When the chain is 5 feet high it still has a positive velocity of about 7.3 ft/s, which keeps it going higher for a while.

18. (a) There are two forces acting on the chain as it falls from the platform. One is the force due to gravity on the portion of the chain hanging over the edge of the platform. This is $F_1 = 2x$. The second is due to the motion of the portion of the chain stretched out on the platform. By

Newton's second law this is

$$F_2 = \frac{d}{dt}[mv] = \frac{d}{dt}\left[\frac{(8-x)2}{32}v\right] = \frac{d}{dt}\left[\frac{8-x}{16}v\right]$$

$$= \frac{8-x}{16}\frac{dv}{dt} - \frac{1}{16}v\frac{dx}{dt} = \frac{1}{16}\left[(8-x)\frac{dv}{dt} - v^2\right].$$

From $\frac{d}{dt}[mv] = F_1 - F_2$ we have

$$\frac{d}{dt}\left[\frac{2x}{32}v\right] = 2x - \frac{1}{16}\left[(8-x)\frac{dv}{dt} - v^2\right]$$

$$\frac{x}{16}\frac{dv}{dt} + \frac{1}{16}v\frac{dx}{dt} = 2x - \frac{1}{16}\left[(8-x)\frac{dv}{dt} - v^2\right]$$

$$x\frac{dv}{dt} + v^2 = 32x - (8-x)\frac{dv}{dt} + v^2$$

$$x\frac{dv}{dt} = 32x - 8\frac{dv}{dt} + x\frac{dv}{dt}$$

$$8\frac{dv}{dt} = 32x.$$

By the chain rule, $dv/dt = (dv/dx)(dx/dt) = v\,dv/dx$, so

$$8\frac{dv}{dt} = 8v\frac{dv}{dx} = 32x \quad \text{and} \quad v\frac{dv}{dx} = 4x.$$

(b) Integrating $v\,dv = 4x\,dx$ we get $\frac{1}{2}v^2 = 2x^2 + c$. Since $v = 0$ when $x = 3$, we have $c = -18$. Then $v^2 = 4x^2 - 36$ and $v = \sqrt{4x^2 - 36}$. Using $v = dx/dt$, separating variables, and integrating we obtain

$$\frac{dx}{\sqrt{x^2 - 9}} = 2\,dt \quad \text{and} \quad \cosh^{-1}\frac{x}{3} = 2t + c_1.$$

Solving for x we get $x(t) = 3\cosh(2t + c_1)$. Since $x = 3$ when $t = 0$, we have $\cosh c_1 = 1$ and $c_1 = 0$. Thus, $x(t) = 3\cosh 2t$. Differentiating, we find $v(t) = dx/dt = 6\sinh 2t$.

(c) To find when the back end of the chain will leave the platform we solve $x(t) = 3\cosh 2t = 8$. This gives $t_1 = \frac{1}{2}\cosh^{-1}\frac{8}{3} \approx 0.8184$ seconds. The velocity at this instant is $v(t_1) = 6\sinh(\cosh^{-1}\frac{8}{3}) = 2\sqrt{55} \approx 14.83$ ft/s.

(d) Replacing 8 with L and 32 with g in part (a) we have $L\,dv/dt = gx$. Then

$$L\frac{dv}{dt} = Lv\frac{dv}{dx} = gx \quad \text{and} \quad v\frac{dv}{dx} = \frac{g}{L}x.$$

Integrating we get $\frac{1}{2}v^2 = \frac{g}{2L}x^2 + c$. Setting $x = x_0$ and $v = 0$, we find $c = -\frac{g}{2L}x_0^2$. Solving for v we find

$$v(x) = \sqrt{\frac{g}{L}x^2 - \frac{g}{L}x_0^2}\,.$$

Then the velocity at which the end of the chain leaves the edge of the platform is

$$v(L) = \sqrt{\frac{g}{L}(L^2 - x_0^2)} \ .$$

——————— Chapter 5 Review Exercises ———————

3. $5/4$ m, since $x = -\cos 4t + \frac{3}{4}\sin 4t$.

6. False

9. The period of a spring mass system is given by $T = 2\pi/\omega$ where $\omega^2 = k/m = kg/W$, where k is the spring constant, W is the weight of the mass attached to the spring, and g is the acceleration due to gravity. Thus, the period of oscillation is $T = (2\pi/\sqrt{kg})\sqrt{W}$. If the weight of the original mass is W, then $(2\pi/\sqrt{kg})\sqrt{W} = 3$ and $(2\pi/\sqrt{kg})\sqrt{W-8} = 2$. Dividing, we get $\sqrt{W}/\sqrt{W-8} = 3/2$ or $W = \frac{9}{4}(W-8)$. Solving for W we find that the weight of the original mass was 14.4 pounds.

12. From $x'' + \beta x' + 64x = 0$ we see that oscillatory motion results if $\beta^2 - 256 < 0$ or $0 \le |\beta| < 16$.

15. Writing $\frac{1}{8}x'' + \frac{8}{3}x = \cos\gamma t + \sin\gamma t$ in the form $x'' + \frac{64}{3}x = 8\cos\gamma t + 8\sin\gamma t$ we identify $\lambda = 0$ and $\omega^2 = 64/3$. The system is in a state of pure resonance when $\gamma = \omega = \sqrt{64/3} = 8/\sqrt{3}$.

18. **(a)** Let k be the effective spring constant and x_1 and x_2 the elongation of springs k_1 and k_2. The restoring forces satisfy $k_1 x_1 = k_2 x_2$ so $x_2 = (k_1/k_2)x_1$. From $k(x_1 + x_2) = k_1 x_1$ we have

$$k\left(x_1 + \frac{k_1}{k_2}x_2\right) = k_1 x_1$$

$$k\left(\frac{k_2 + k_1}{k_2}\right) = k_1$$

$$k = \frac{k_1 k_2}{k_1 + k_2}$$

$$\frac{1}{k} = \frac{1}{k_1} + \frac{1}{k_2} \ .$$

(b) From $k_1 = 2W$ and $k_2 = 4W$ we find $1/k = 1/2W + 1/4W = 3/4W$. Then $k = 4W/3 = 4mg/3$. The differential equation $mx'' + kx = 0$ then becomes $x'' + (4g/3)x = 0$. The solution is

$$x(t) = c_1 \cos 2\sqrt{\frac{g}{3}}\,t + c_2 \sin 2\sqrt{\frac{g}{3}}\,t.$$

The initial conditions $x(0) = 1$ and $x'(0) = 2/3$ imply $c_1 = 1$ and $c_2 = 1/\sqrt{3g}$.

(c) To compute the maximum speed of the weight we compute

$$x'(t) = 2\sqrt{\frac{g}{3}}\sin 2\sqrt{\frac{g}{3}}\,t + \frac{2}{3}\cos 2\sqrt{\frac{g}{3}}\,t \quad \text{and} \quad |x'(t)| = \sqrt{4\frac{g}{3} + \frac{4}{9}} = \frac{2}{3}\sqrt{3g+1}\ .$$

21. For $\lambda > 0$ the general solution is $y = c_1 \cos \sqrt{\lambda} x + c_2 \sin \sqrt{\lambda} x$. Now $y(0) = c_1$ and $y(2\pi) = c_1 \cos 2\pi \sqrt{\lambda} + c_2 \sin 2\pi \sqrt{\lambda}$, so the condition $y(0) = y(2\pi)$ implies

$$c_1 = c_1 \cos 2\pi \sqrt{\lambda} + c_2 \sin 2\pi \sqrt{\lambda}$$

which is true when $\sqrt{\lambda} = n$ or $\lambda = n^2$ for $n = 1, 2, 3, \ldots$. Since

$$y' = -\sqrt{\lambda}\, c_1 \sin \sqrt{\lambda} x + \sqrt{\lambda}\, c_2 \cos \sqrt{\lambda} x = -nc_1 \sin nx + nc_2 \cos nx,$$

we see that $y'(0) = nc_2 = y'(2\pi)$ for $n = 1, 2, 3, \ldots$. Thus, the eigenvalues are n^2 for $n = 1, 2, 3, \ldots$, with corresponding eigenfunctions $\cos nx$ and $\sin nx$. When $\lambda = 0$, the general solution is $y = c_1 x + c_2$ and the corresponding eigenfunction is $y = 1$.

For $\lambda < 0$ the general solution is $y = c_1 \cosh \sqrt{-\lambda} x + c_2 \sinh \sqrt{-\lambda} x$. In this case $y(0) = c_1$ and $y(2\pi) = c_1 \cosh 2\pi \sqrt{-\lambda} + c_2 \sinh 2\pi \sqrt{-\lambda}$, so $y(0) = y(2\pi)$ can only be valid for $\lambda = 0$. Thus, there are no eigenvalues corresponding to $\lambda < 0$.

6 Series Solutions of Linear Equations

—————— **Exercises 6.1** ——————————

3. $\lim\limits_{k\to\infty}\left|\dfrac{a_{k+1}}{a_k}\right| = \lim\limits_{k\to\infty}\left|\dfrac{(x-5)^{k+1}/10^{k+1}}{(x-5)^k/10^k}\right| = \lim\limits_{k\to\infty}\dfrac{1}{10}|x-5| = \dfrac{1}{10}|x-5|$

The series is absolutely convergent for $\frac{1}{10}|x-5| < 1$, $|x-5| < 10$, or on $(-5, 15)$. At $x = -5$, the series $\sum\limits_{k=1}^{\infty}\dfrac{(-1)^k(-10)^k}{10^k} = \sum\limits_{k=1}^{\infty} 1$ diverges by the k-th term test. At $x = 15$, the series $\sum\limits_{k=1}^{\infty}\dfrac{(-1)^k 10^k}{10^k} = \sum\limits_{k=1}^{\infty}(-1)^k$ diverges by the k-th term test. Thus, the series converges on $(-5, 15)$.

6. $e^{-x}\cos x = \left(1 - x + \dfrac{x^2}{2} - \dfrac{x^3}{6} + \dfrac{x^4}{24} - \cdots\right)\left(1 - \dfrac{x^2}{2} + \dfrac{x^4}{24} - \cdots\right) = 1 - x + \dfrac{x^3}{3} - \dfrac{x^4}{6} + \cdots$

9. $\sum\limits_{n=1}^{\infty} 2nc_n x^{n-1} + \sum\limits_{n=0}^{\infty} 6c_n x^{n+1} = 2\cdot 1 \cdot c_1 x^0 + \underbrace{\sum\limits_{n=2}^{\infty} 2nc_n x^{n-1}}_{k=n-1} + \underbrace{\sum\limits_{n=0}^{\infty} 6c_n x^{n+1}}_{k=n+1}$

$$= 2c_1 + \sum\limits_{k=1}^{\infty} 2(k+1)c_{k+1}x^k + \sum\limits_{k=1}^{\infty} 6c_{k-1}x^k$$

$$= 2c_1 + \sum\limits_{k=1}^{\infty}[2(k+1)c_{k+1} + 6c_{k-1}]x^k$$

12. $y' = \sum\limits_{n=1}^{\infty}\dfrac{(-1)^n 2n}{2^{2n}(n!)^2}x^{2n-1}$, $\qquad y'' = \sum\limits_{n=1}^{\infty}\dfrac{(-1)^n 2n(2n-1)}{2^{2n}(n!)^2}x^{2n-2}$

$xy'' + y' + xy = \underbrace{\sum\limits_{n=1}^{\infty}\dfrac{(-1)^n 2n(2n-1)}{2^{2n}(n!)^2}x^{2n-1}}_{k=n} + \underbrace{\sum\limits_{n=1}^{\infty}\dfrac{(-1)^n 2n}{2^{2n}(n!)^2}x^{2n-1}}_{k=n} + \underbrace{\sum\limits_{n=0}^{\infty}\dfrac{(-1)^n}{2^{2n}(n!)^2}x^{2n+1}}_{k=n+1}$

$$= \sum\limits_{k=1}^{\infty}\left[\dfrac{(-1)^k 2k(2k-1)}{2^{2k}(k!)^2} + \dfrac{(-1)^k 2k}{2^{2k}(k!)^2} + \dfrac{(-1)^{k-1}}{2^{2k-2}[(k-1)!]^2}\right]x^{2k-1}$$

$$= \sum\limits_{k=1}^{\infty}\left[\dfrac{(-1)^k(2k)^2}{2^{2k}(k!)^2} - \dfrac{(-1)^k}{2^{2k-2}[(k-1)!]^2}\right]x^{2k-1}$$

$$= \sum\limits_{k=1}^{\infty}(-1)^k\left[\dfrac{(2k)^2 - 2^2 k^2}{2^{2k}(k!)^2}\right]x^{2k-1} = 0$$

64

15. Substituting $y = \sum_{n=0}^{\infty} c_n x^n$ into the differential equation we have

$$y'' - 2xy' + y = \underbrace{\sum_{n=2}^{\infty} n(n-1)c_n x^{n-2}}_{k=n-2} - 2\underbrace{\sum_{n=1}^{\infty} nc_n x^n}_{k=n} + \underbrace{\sum_{n=0}^{\infty} c_n x^n}_{k=n}$$

$$= \sum_{k=0}^{\infty} (k+2)(k+1)c_{k+2}x^k - 2\sum_{k=1}^{\infty} kc_k x^k + \sum_{k=0}^{\infty} c_k x^k$$

$$= 2c_2 + c_0 + \sum_{k=1}^{\infty} [(k+2)(k+1)c_{k+2} - (2k-1)c_k]x^k = 0.$$

Thus

$$2c_2 + c_0 = 0$$

$$(k+2)(k+1)c_{k+2} - (2k-1)c_k = 0$$

and

$$c_2 = -\frac{1}{2}c_0$$

$$c_{k+2} = \frac{2k-1}{(k+2)(k+1)}c_k, \quad k = 1, 2, 3, \dots.$$

Choosing $c_0 = 1$ and $c_1 = 0$ we find

$$c_2 = -\frac{1}{2}$$

$$c_3 = c_5 = c_7 = \cdots = 0$$

$$c_4 = -\frac{1}{8}$$

$$c_6 = -\frac{7}{336}$$

and so on. For $c_0 = 0$ and $c_1 = 1$ we obtain

$$c_2 = c_4 = c_6 = \cdots = 0$$

$$c_3 = \frac{1}{6}$$

$$c_5 = \frac{1}{24}$$

$$c_7 = \frac{1}{112}$$

and so on. Thus, two solutions are

$$y_1 = 1 - \frac{1}{2}x^2 - \frac{1}{8}x^4 - \frac{7}{336}x^6 - \cdots \qquad \text{and} \qquad y_2 = x + \frac{1}{6}x^3 + \frac{1}{24}x^5 + \frac{1}{112}x^7 + \cdots.$$

18. Substituting $y = \sum_{n=0}^{\infty} c_n x^n$ into the differential equation we have

$$y'' + 2xy' + 2y = \underbrace{\sum_{n=2}^{\infty} n(n-1)c_n x^{n-2}}_{k=n-2} + 2\underbrace{\sum_{n=1}^{\infty} nc_n x^n}_{k=n} + 2\underbrace{\sum_{n=0}^{\infty} c_n x^n}_{k=n}$$

$$= \sum_{k=0}^{\infty} (k+2)(k+1)c_{k+2}x^k + 2\sum_{k=1}^{\infty} kc_k x^k + 2\sum_{k=0}^{\infty} c_k x^k$$

$$= 2c_2 + 2c_0 + \sum_{k=1}^{\infty} [(k+2)(k+1)c_{k+2} + 2(k+1)c_k]x^k = 0.$$

Thus

$$2c_2 + 2c_0 = 0$$

$$(k+2)(k+1)c_{k+2} + 2(k+1)c_k = 0$$

and

$$c_2 = -c_0$$

$$c_{k+2} = -\frac{2}{k+2}c_k, \quad k = 1, 2, 3, \ldots.$$

Choosing $c_0 = 1$ and $c_1 = 0$ we find

$$c_2 = -1$$

$$c_3 = c_5 = c_7 = \cdots = 0$$

$$c_4 = \frac{1}{2}$$

$$c_6 = -\frac{1}{6}$$

and so on. For $c_0 = 0$ and $c_1 = 1$ we obtain

$$c_2 = c_4 = c_6 = \cdots = 0$$

$$c_3 = -\frac{2}{3}$$

$$c_5 = \frac{4}{15}$$

$$c_7 = -\frac{8}{105}$$

and so on. Thus, two solutions are

$$y_1 = 1 - x^2 + \frac{1}{2}x^4 - \frac{1}{6}x^6 + \cdots \quad \text{and} \quad y_2 = x - \frac{2}{3}x^3 + \frac{4}{15}x^5 - \frac{8}{105}x^7 + \cdots.$$

21. Substituting $y = \sum_{n=0}^{\infty} c_n x^n$ into the differential equation we have

$$y'' - (x+1)y' - y = \underbrace{\sum_{n=2}^{\infty} n(n-1)c_n x^{n-2}}_{k=n-2} - \underbrace{\sum_{n=1}^{\infty} nc_n x^n}_{k=n} - \underbrace{\sum_{n=1}^{\infty} nc_n x^{n-1}}_{k=n-1} - \underbrace{\sum_{n=0}^{\infty} c_n x^n}_{k=n}$$

$$= \sum_{k=0}^{\infty} (k+2)(k+1)c_{k+2} x^k - \sum_{k=1}^{\infty} kc_k x^k - \sum_{k=0}^{\infty} (k+1)c_{k+1} x^k - \sum_{k=0}^{\infty} c_k x^k$$

$$= 2c_2 - c_1 - c_0 + \sum_{k=1}^{\infty} [(k+2)(k+1)c_{k+2} - (k+1)c_{k+1} - (k+1)c_k] x^k = 0.$$

Thus

$$2c_2 - c_1 - c_0 = 0$$

$$(k+2)(k+1)c_{k+2} - (k-1)(c_{k+1} + c_k) = 0$$

and

$$c_2 = \frac{c_1 + c_0}{2}$$

$$c_{k+2} = \frac{c_{k+1} + c_k}{k+2} c_k, \quad k = 2, 3, 4, \ldots .$$

Choosing $c_0 = 1$ and $c_1 = 0$ we find

$$c_2 = \frac{1}{2}, \qquad c_3 = \frac{1}{6}, \qquad c_4 = \frac{1}{6}$$

and so on. For $c_0 = 0$ and $c_1 = 1$ we obtain

$$c_2 = \frac{1}{2}, \qquad c_3 = \frac{1}{2}, \qquad c_4 = \frac{1}{4}$$

and so on. Thus, two solutions are

$$y_1 = 1 + \frac{1}{2}x^2 + \frac{1}{6}x^3 + \frac{1}{6}x^4 + \cdots \qquad \text{and} \qquad y_2 = x + \frac{1}{2}x^2 + \frac{1}{2}x^3 + \frac{1}{4}x^4 + \cdots .$$

24. Substituting $y = \sum_{n=0}^{\infty} c_n x^n$ into the differential equation we have

$$\left(x^2 - 1\right) y'' + xy' - y = \underbrace{\sum_{n=2}^{\infty} n(n-1)c_n x^n}_{k=n} - \underbrace{\sum_{n=2}^{\infty} n(n-1)c_n x^{n-2}}_{k=n-2} + \underbrace{\sum_{n=1}^{\infty} nc_n x^n}_{k=n} - \underbrace{\sum_{n=0}^{\infty} c_n x^n}_{k=n}$$

$$= \sum_{k=2}^{\infty} k(k-1)c_k x^k - \sum_{k=0}^{\infty} (k+2)(k+1)c_{k+2} x^k + \sum_{k=1}^{\infty} kc_k x^k - \sum_{k=0}^{\infty} c_k x^k$$

$$= (-2c_2 - c_0) - 6c_3 x + \sum_{k=2}^{\infty} \left[-(k+2)(k+1)c_{k+2} + \left(k^2 - 1\right) c_k \right] x^k = 0.$$

Thus

$$-2c_2 - c_0 = 0$$

$$-6c_3 = 0$$

$$-(k+2)(k+1)c_{k+2} + (k-1)(k+1)c_k = 0$$

and

$$c_2 = -\frac{1}{2}c_0$$

$$c_3 = 0$$

$$c_{k+2} = \frac{k-1}{k+2}c_k, \quad k = 2, 3, 4, \ldots .$$

Choosing $c_0 = 1$ and $c_1 = 0$ we find

$$c_2 = -\frac{1}{2}$$

$$c_3 = c_5 = c_7 = \cdots = 0$$

$$c_4 = -\frac{1}{8}$$

and so on. For $c_0 = 0$ and $c_1 = 1$ we obtain

$$c_2 = c_4 = c_6 = \cdots = 0$$

$$c_3 = c_5 = c_7 = \cdots = 0.$$

Thus, two solutions are

$$y_1 = 1 - \frac{1}{2}x^2 - \frac{1}{8}x^4 - \cdots \qquad \text{and} \qquad y_2 = x.$$

27. Substituting $y = \sum_{n=0}^{\infty} c_n x^n$ into the differential equation we have

$$y'' - 2xy' + 8y = \underbrace{\sum_{n=2}^{\infty} n(n-1)c_n x^{n-2}}_{k=n-2} - 2\underbrace{\sum_{n=1}^{\infty} nc_n x^n}_{k=n} + 8\underbrace{\sum_{n=0}^{\infty} c_n x^n}_{k=n}$$

$$= \sum_{k=0}^{\infty} (k+2)(k+1)c_{k+2}x^k - 2\sum_{k=1}^{\infty} kc_k x^k + 8\sum_{k=0}^{\infty} c_k x^k$$

$$= 2c_2 + 8c_0 + \sum_{k=1}^{\infty} [(k+2)(k+1)c_{k+2} + (8-2k)c_k]x^k = 0.$$

Thus

$$2c_2 + 8c_0 = 0$$

68

$$(k+2)(k+1)c_{k+2} + (8-2k)c_k = 0$$

and

$$c_2 = -4c_0$$

$$c_{k+2} = \frac{2k-8}{(k+2)(k+1)}c_k, \quad k = 1,2,3,\ldots.$$

Choosing $c_0 = 1$ and $c_1 = 0$ we find

$$c_2 = -4$$

$$c_3 = c_5 = c_7 = \cdots = 0$$

$$c_4 = \frac{4}{3}$$

$$c_6 = c_8 = c_{10} = \cdots = 0.$$

For $c_0 = 0$ and $c_1 = 1$ we obtain

$$c_2 = c_4 = c_6 = \cdots = 0$$

$$c_3 = -1$$

$$c_5 = \frac{1}{10}$$

and so on. Thus,

$$y = C_1\left(1 - 4x^2 + \frac{4}{3}x^4\right) + C_2\left(x - x^3 + \frac{1}{10}x^5 + \cdots\right)$$

and

$$y' = C_1\left(-8x + \frac{16}{3}x^3\right) + C_2\left(1 - 3x^2 + \frac{1}{2}x^4 + \cdots\right).$$

The initial conditions imply $C_1 = 3$ and $C_2 = 0$, so

$$y = 3\left(1 - 4x^2 + \frac{4}{3}x^4\right) = 3 - 12x^2 + 4x^4.$$

30. Substituting $y = \sum_{n=0}^{\infty} c_n x^n$ into the differential equation we have

$$y'' + e^x y' - y = \sum_{n=2}^{\infty} n(n-1)c_n x^{n-2}$$

$$+ \left(1 + x + \frac{1}{2}x^2 + \frac{1}{6}x^3 + \cdots\right)\left(c_1 + 2c_2 x + 3c_3 x^2 + 4c_4 x^3 + \cdots\right) - \sum_{n=0}^{\infty} c_n x^n$$

$$= \left[2c_2 + 6c_3 x + 12c_4 x^2 + 20c_5 x^3 + \cdots\right]$$

$$+ \left[c_1 + (2c_2 + c_1)x + \left(3c_3 + 2c_2 + \frac{1}{2}c_1\right)x^2 + \cdots\right] - [c_0 + c_1 x + c_2 x^2 + \cdots]$$

$$= (2c_2 + c_1 - c_0) + (6c_3 + 2c_2)x + \left(12c_4 + 3c_3 + c_2 + \frac{1}{2}c_1\right)x^2 + \cdots = 0.$$

Thus

$$2c_2 + c_1 - c_0 = 0$$

$$6c_3 + 2c_2 = 0$$

$$12c_4 + 3c_3 + c_2 + \frac{1}{2}c_1 = 0$$

and

$$c_2 = \frac{1}{2}c_0 - \frac{1}{2}c_1$$

$$c_3 = -\frac{1}{3}c_2$$

$$c_4 = -\frac{1}{4}c_3 + \frac{1}{12}c_2 - \frac{1}{24}c_1.$$

Choosing $c_0 = 1$ and $c_1 = 0$ we find

$$c_2 = \frac{1}{2}, \qquad c_3 = -\frac{1}{6}, \qquad c_4 = 0$$

and so on. For $c_0 = 0$ and $c_1 = 1$ we obtain

$$c_2 = -\frac{1}{2}, \qquad c_3 = \frac{1}{6}, \qquad c_4 = -\frac{1}{24}$$

and so on. Thus, two solutions are

$$y_1 = 1 + \frac{1}{2}x^2 - \frac{1}{6}x^3 + \cdots \qquad \text{and} \qquad y_2 = x - \frac{1}{2}x^2 + \frac{1}{6}x^3 - \frac{1}{24}x^4 + \cdots.$$

3. Irregular singular point: $x = 3$; regular singular point: $x = -3$

6. Irregular singular point: $x = 5$; regular singular point: $x = 0$

9. Irregular singular point: $x = 0$; regular singular points: $x = 2, \pm 5$

12. Writing the differential equation in the form

$$y'' + \frac{x+3}{x}y' + 7xy = 0$$

we see that $x_0 = 0$ is a regular singular point. Multiplying by x^2, the differential equation can be put in the form

$$x^2 y'' + x(x+3)y' + 7x^3 y = 0.$$

We identify $p(x) = x + 3$ and $q(x) = 7x^3$.

15. Substituting $y = \sum_{n=0}^{\infty} c_n x^{n+r}$ into the differential equation and collecting terms, we obtain

$$2xy'' - y' + 2y = \left(2r^2 - 3r\right) c_0 x^{r-1} + \sum_{k=1}^{\infty} [2(k+r-1)(k+r)c_k - (k+r)c_k + 2c_{k-1}]x^{k+r-1} = 0,$$

which implies

$$2r^2 - 3r = r(2r - 3) = 0$$

and

$$(k+r)(2k+2r-3)c_k + 2c_{k-1} = 0.$$

The indicial roots are $r = 0$ and $r = 3/2$. For $r = 0$ the recurrence relation is

$$c_k = -\frac{2c_{k-1}}{k(2k-3)}, \quad k = 1, 2, 3, \ldots,$$

and

$$c_1 = 2c_0, \qquad c_2 = -2c_0, \qquad c_3 = \frac{4}{9}c_0.$$

For $r = 3/2$ the recurrence relation is

$$c_k = -\frac{2c_{k-1}}{(2k+3)k}, \quad k = 1, 2, 3, \ldots,$$

and

$$c_1 = -\frac{2}{5}c_0, \qquad c_2 = \frac{2}{35}c_0, \qquad c_3 = -\frac{4}{945}c_0.$$

The general solution on $(0, \infty)$ is

$$y = C_1 \left(1 + 2x - 2x^2 + \frac{4}{9}x^3 + \cdots\right) + C_2 x^{3/2}\left(1 - \frac{2}{5}x + \frac{2}{35}x^2 - \frac{4}{945}x^3 + \cdots\right).$$

18. Substituting $y = \sum_{n=0}^{\infty} c_n x^{n+r}$ into the differential equation and collecting terms, we obtain

$$2x^2 y'' - xy' + \left(x^2 + 1\right) y = \left(2r^2 - 3r + 1\right) c_0 x^r + \left(2r^2 + r\right) c_1 x^{r+1}$$

$$+ \sum_{k=2}^{\infty} [2(k+r)(k+r-1)c_k - (k+r)c_k + c_k + c_{k-2}]x^{k+r}$$

$$= 0,$$

which implies

$$2r^2 - 3r + 1 = (2r-1)(r-1) = 0,$$

$$\left(2r^2 + r\right) c_1 = 0,$$

and

$$[(k+r)(2k+2r-3) + 1]c_k + c_{k-2} = 0.$$

The indicial roots are $r = 1/2$ and $r = 1$, so $c_1 = 0$. For $r = 1/2$ the recurrence relation is

$$c_k = -\frac{c_{k-2}}{k(2k-1)}, \quad k = 2, 3, 4, \ldots,$$

and

$$c_2 = -\frac{1}{6}c_0, \qquad c_3 = 0, \qquad c_4 = \frac{1}{168}c_0.$$

For $r = 1$ the recurrence relation is

$$c_k = -\frac{c_{k-2}}{k(2k+1)}, \quad k = 2, 3, 4, \ldots,$$

and

$$c_2 = -\frac{1}{10}c_0, \qquad c_3 = 0, \qquad c_4 = \frac{1}{360}c_0.$$

The general solution on $(0, \infty)$ is

$$y = C_1 x^{1/2} \left(1 - \frac{1}{6}x^2 + \frac{1}{168}x^4 + \cdots\right) + C_2 x \left(1 - \frac{1}{10}x^2 + \frac{1}{360}x^4 + \cdots\right).$$

21. Substituting $y = \sum_{n=0}^{\infty} c_n x^{n+r}$ into the differential equation and collecting terms, we obtain

$$2xy'' - (3 + 2x)y' + y = \left(2r^2 - 5r\right) c_0 x^{r-1} + \sum_{k=1}^{\infty} [2(k+r)(k+r-1)c_k$$

$$- 3(k+r)c_k - 2(k+r-1)c_{k-1} + c_{k-1}]x^{k+r-1}$$

$$= 0,$$

which implies

$$2r^2 - 5r = r(2r - 5) = 0$$

and

$$(k + r)(2k + 2r - 5)c_k - (2k + 2r - 3)c_{k-1} = 0.$$

The indicial roots are $r = 0$ and $r = 5/2$. For $r = 0$ the recurrence relation is

$$c_k = \frac{(2k - 3)c_{k-1}}{k(2k - 5)}, \quad k = 1, 2, 3, \ldots,$$

and

$$c_1 = \frac{1}{3}c_0, \qquad c_2 = -\frac{1}{6}c_0, \qquad c_3 = -\frac{1}{6}c_0.$$

For $r = 5/2$ the recurrence relation is

$$c_k = \frac{2(k + 1)c_{k-1}}{k(2k + 5)}, \quad k = 1, 2, 3, \ldots,$$

and

$$c_1 = \frac{4}{7}c_0, \qquad c_2 = \frac{4}{21}c_0, \qquad c_3 = \frac{32}{693}c_0.$$

The general solution on $(0, \infty)$ is

$$y = C_1 \left(1 + \frac{1}{3}x - \frac{1}{6}x^2 - \frac{1}{6}x^3 + \cdots\right) + C_2 x^{5/2}\left(1 + \frac{4}{7}x + \frac{4}{21}x^2 + \frac{32}{693}x^3 + \cdots\right).$$

24. Substituting $y = \sum_{n=0}^{\infty} c_n x^{n+r}$ into the differential equation and collecting terms, we obtain

$$2x^2 y'' + 3xy' + (2x - 1)y = \left(2r^2 + r - 1\right)c_0 x^r$$

$$+ \sum_{k=1}^{\infty} [2(k + r)(k + r - 1)c_k + 3(k + r)c_k - c_k + 2c_{k-1}]x^{k+r}$$

$$= 0,$$

which implies

$$2r^2 + r - 1 = (2r - 1)(r + 1) = 0$$

and

$$[(k + r)(2k + 2r + 1) - 1]c_k + 2c_{k-1} = 0.$$

The indicial roots are $r = -1$ and $r = 1/2$. For $r = -1$ the recurrence relation is

$$c_k = -\frac{2c_{k-1}}{k(2k - 3)}, \quad k = 1, 2, 3, \ldots,$$

and

$$c_1 = 2c_0, \qquad c_2 = -2c_0, \qquad c_3 = \frac{4}{9}c_0.$$

For $r = 1/2$ the recurrence relation is

$$c_k = -\frac{2c_{k-1}}{k(2k + 3)}, \quad k = 1, 2, 3, \ldots,$$

73

and

$$c_1 = -\frac{2}{5}c_0, \qquad c_2 = \frac{2}{35}c_0, \qquad c_3 = -\frac{4}{945}c_0.$$

The general solution on $(0, \infty)$ is

$$y = C_1 x^{-1}\left(1 + 2x - 2x^2 + \frac{4}{9}x^3 + \cdots\right) + C_2 x^{1/2}\left(1 - \frac{2}{5}x + \frac{2}{35}x^2 - \frac{4}{945}x^3 + \cdots\right).$$

27. Substituting $y = \sum_{n=0}^{\infty} c_n x^{n+r}$ into the differential equation and collecting terms, we obtain

$$xy'' - xy' + y = \left(r^2 - r\right)c_0 x^{r-1} + \sum_{k=0}^{\infty}[(k+r+1)(k+r)c_{k+1} - (k+r)c_k + c_k]x^{k+r} = 0$$

which implies

$$r^2 - r = r(r-1) = 0$$

and

$$(k+r+1)(k+r)c_{k+1} - (k+r-1)c_k = 0.$$

The indicial roots are $r_1 = 1$ and $r_2 = 0$. For $r_1 = 1$ the recurrence relation is

$$c_{k+1} = \frac{kc_k}{(k+2)(k+1)}, \qquad k = 0, 1, 2, \ldots,$$

and one solution is $y_1 = c_0 x$. A second solution is

$$y_2 = x \int \frac{e^{-\int -dx}}{x^2}\, dx = x \int \frac{e^x}{x^2}\, dx = x \int \frac{1}{x^2}\left(1 + x + \frac{1}{2}x^2 + \frac{1}{3!}x^3 + \cdots\right) dx$$

$$= x \int \left(\frac{1}{x^2} + \frac{1}{x} + \frac{1}{2} + \frac{1}{3!}x + \frac{1}{4!}x^2 + \cdots\right) dx = x\left[-\frac{1}{x} + \ln x + \frac{1}{2}x + \frac{1}{12}x^2 + \frac{1}{72}x^3 + \cdots\right]$$

$$= x \ln x - 1 + \frac{1}{2}x^2 + \frac{1}{12}x^3 + \frac{1}{72}x^4 + \cdots.$$

The general solution on $(0, \infty)$ is

$$y = C_1 x + C_2 y_2(x).$$

30. Substituting $y = \sum_{n=0}^{\infty} c_n x^{n+r}$ into the differential equation and collecting terms, we obtain

$$xy'' + y' + y = r^2 c_0 x^{r-1} + \sum_{k=1}^{\infty}[(k+r)(k+r-1)c_k + (k+r)c_k + c_{k-1}]x^{k+r-1} = 0$$

which implies $r^2 = 0$ and

$$(k+r)^2 c_k + c_{k-1} = 0.$$

The indicial roots are $r_1 = r_2 = 0$ and the recurrence relation is

$$c_k = -\frac{c_{k-1}}{k^2}, \qquad k = 1, 2, 3, \ldots.$$

74

One solution is

$$y_1 = c_0\left(1 - x + \frac{1}{2^2}x^2 - \frac{1}{(3!)^2}x^3 + \frac{1}{(4!)^2}x^4 - \cdots\right) = c_0\sum_{n=0}^{\infty}\frac{(-1)^n}{(n!)^2}x^n.$$

A second solution is

$$y_2 = y_1\int\frac{e^{-\int(1/x)dx}}{y_1^2}\,dx = y_1\int\frac{dx}{x\left(1 - x + \frac{1}{4}x^2 - \frac{1}{36}x^3 + \cdots\right)^2}$$

$$= y_1\int\frac{dx}{x\left(1 - 2x + \frac{3}{2}x^2 - \frac{5}{9}x^3 + \frac{35}{288}x^4 - \cdots\right)}$$

$$= y_1\int\frac{1}{x}\left(1 + 2x + \frac{5}{2}x^2 + \frac{23}{9}x^3 + \frac{677}{288}x^4 + \cdots\right)dx$$

$$= y_1\int\left(\frac{1}{x} + 2 + \frac{5}{2}x + \frac{23}{9}x^2 + \frac{677}{288}x^3 + \cdots\right)dx$$

$$= y_1\left[\ln x + 2x + \frac{5}{4}x^2 + \frac{23}{27}x^3 + \frac{677}{1,152}x^4 + \cdots\right]$$

$$= y_1\ln x + y_1\left(2x + \frac{5}{4}x^2 + \frac{23}{27}x^3 + \frac{677}{1,152}x^4 + \cdots\right).$$

The general solution on $(0,\infty)$ is

$$y = C_1 y_1(x) + C_2 y_2(x).$$

33. (a) From $t = 1/x$ we have $dt/dx = -1/x^2 = -t^2$. Then

$$\frac{dy}{dx} = \frac{dy}{dt}\frac{dt}{dx} = -t^2\frac{dy}{dt}$$

and

$$\frac{d^2y}{dx^2} = \frac{d}{dx}\left(\frac{dy}{dx}\right) = \frac{d}{dx}\left(-t^2\frac{dy}{dt}\right) = -t^2\frac{d^2y}{dt^2}\frac{dt}{dx} - \frac{dy}{dt}\left(2t\frac{dt}{dx}\right) = t^4\frac{d^2y}{dt^2} + 2t^3\frac{dy}{dt}.$$

Now

$$x^4\frac{d^2y}{dx^2} + \lambda y = \frac{1}{t^4}\left(t^4\frac{d^2y}{dt^2} + 2t^3\frac{dy}{dt}\right) + \lambda y = \frac{d^2y}{dt^2} + \frac{2}{t}\frac{dy}{dt} + \lambda y = 0$$

becomes

$$t\frac{d^2y}{dt^2} + 2\frac{dy}{dt} + \lambda t y = 0.$$

75

(b) Substituting $y = \sum_{n=0}^{\infty} c_n t^{n+r}$ into the differential equation and collecting terms, we obtain

$$t\frac{d^2y}{dt^2} + 2\frac{dy}{dt} + \lambda t y = (r^2 + r)c_0 t^{r-1} + (r^2 + 3r + 2)c_1 t^r$$

$$+ \sum_{k=2}^{\infty} [(k+r)(k+r-1)c_k + 2(k+r)c_k + \lambda c_{k-2}]t^{k+r-1}$$

$$= 0,$$

which implies

$$r^2 + r = r(r+1) = 0,$$

$$\left(r^2 + 3r + 2\right)c_1 = 0,$$

and

$$(k+r)(k+r+1)c_k + \lambda c_{k-2} = 0.$$

The indicial roots are $r_1 = 0$ and $r_2 = -1$, so $c_1 = 0$. For $r_1 = 0$ the recurrence relation is

$$c_k = -\frac{\lambda c_{k-2}}{k(k+1)}, \quad k = 2, 3, 4, \ldots,$$

and

$$c_2 = -\frac{\lambda}{3!}c_0$$

$$c_3 = c_5 = c_7 = \cdots = 0$$

$$c_4 = \frac{\lambda^2}{5!}c_0$$

$$c_{2n} = (-1)^n \frac{\lambda^n}{(2n+1)!}c_0.$$

For $r_2 = -1$ the recurrence relation is

$$c_k = -\frac{\lambda c_{k-2}}{k(k-1)}, \quad k = 2, 3, 4, \ldots,$$

and

$$c_2 = -\frac{\lambda}{2!}c_0$$

$$c_3 = c_5 = c_7 = \cdots = 0$$

$$c_4 = \frac{\lambda^2}{4!}c_0$$

$$c_{2n} = (-1)^n \frac{\lambda^n}{(2n)!}c_0.$$

76

The general solution on $(0, \infty)$ is

$$y(t) = C_1 \sum_{n=0}^{\infty} \frac{(-1)^n}{(2n+1)!} (\sqrt{\lambda} t)^{2n} + C_2 t^{-1} \sum_{n=0}^{\infty} \frac{(-1)^n}{(2n)!} (\sqrt{\lambda} t)^{2n}$$

$$= \frac{1}{t} \left[C_1 \sum_{n=0}^{\infty} \frac{(-1)^n}{(2n+1)!} (\sqrt{\lambda} t)^{2n+1} + C_2 \sum_{n=0}^{\infty} \frac{(-1)^n}{(2n)!} (\sqrt{\lambda} t)^{2n} \right]$$

$$= \frac{1}{t} [C_1 \sin \sqrt{\lambda} t + C_2 \cos \sqrt{\lambda} t].$$

(c) Using $t = 1/x$, the solution of the original equation is

$$y(x) C_1 x \sin \frac{\sqrt{\lambda}}{x} + C_2 x \cos \frac{\sqrt{\lambda}}{x}.$$

—— Exercises 6.3 ——

3. Since $\nu^2 = 25/4$ the general solution is $y = c_1 J_{5/2}(x) + c_2 J_{-5/2}(x)$.

6. Since $\nu^2 = 4$ the general solution is $y = c_1 J_2(x) + c_2 Y_2(x)$.

9. If $y = x^{-1/2} v(x)$ then

$$y' = x^{-1/2} v'(x) - \frac{1}{2} x^{-3/2} v(x),$$

$$y'' = x^{-1/2} v''(x) - x^{-3/2} v'(x) + \frac{3}{4} x^{-5/2} v(x),$$

and

$$x^2 y'' + 2x y' + \lambda^2 x^2 y = x^{3/2} v'' + x^{1/2} v' + \left(\lambda^2 x^{3/2} - \frac{1}{4} x^{-1/2} \right) v.$$

Multiplying by $x^{1/2}$ we obtain

$$x^2 v'' + x v' + \left(\lambda^2 x^2 - \frac{1}{4} \right) v = 0,$$

whose solution is $v = c_1 J_{1/2}(\lambda x) + c_2 J_{-1/2}(\lambda x)$. Then $y = c_1 x^{-1/2} J_{1/2}(\lambda x) + c_2 x^{-1/2} J_{-1/2}(\lambda x)$.

12. From $y = \sqrt{x} J_{\nu}(\lambda x)$ we find

$$y' = \lambda \sqrt{x} J_{\nu}'(\lambda x) + \frac{1}{2} x^{-1/2} J_{\nu}(\lambda x)$$

and

$$y'' = \lambda^2 \sqrt{x} J_{\nu}''(\lambda x) + \lambda x^{-1/2} J_{\nu}'(\lambda x) - \frac{1}{4} x^{-3/2} J_{\nu}(\lambda x).$$

Substituting into the differential equation, we have

$$x^2 y'' + \left(\lambda^2 x^2 - \nu^2 + \frac{1}{4}\right) y = \sqrt{x}\left[\lambda^2 x^2 J_\nu''(\lambda x) + \lambda x J_\nu'(\lambda x) + \left(\lambda^2 x^2 - \nu^2\right) J_\nu(\lambda x)\right]$$

$$= \sqrt{x} \cdot 0 \qquad \text{(since } J_n \text{ is a solution of Bessel's equation)}$$

$$= 0.$$

Therefore, $\sqrt{x}\, J_\nu(\lambda x)$ is a solution of the original equation.

15. From Problem 10 with $n = -1$ we find $y = x^{-1} J_{-1}(x)$. From Problem 11 with $n = 1$ we find $y = x^{-1} J_1(x) = -x^{-1} J_{-1}(x)$.

18. From Problem 10 with $n = 3$ we find $y = x^3 J_3(x)$. From Problem 11 with $n = -3$ we find $y = x^3 J_{-3}(x) = -x^3 J_3(x)$.

21. Letting $\nu = 1$ in (15) in the text we have

$$x J_0(x) = \frac{d}{dx}[x J_1(x)] \qquad \text{so} \qquad \int_0^x r J_0(r)\, dr = r J_1(r)\,\Big|_{r=0}^{r=x} = x J_1(x).$$

24. (a) By Problem 20, with $\nu = 1/2$, we obtain $J_{1/2}(x) = x J_{3/2}(x) + x J_{-1/2}(x)$ so that

$$J_{3/2}(x) = \sqrt{\frac{2}{\pi x}}\left(\frac{\sin x}{x} - \cos x\right);$$

with $\nu = -1/2$ we obtain $-J_{-1/2}(x) = x J_{1/2}(x) + x J_{-3/2}(x)$ so that

$$J_{-3/2}(x) = -\sqrt{\frac{2}{\pi x}}\left(\frac{\cos x}{x} + \sin x\right);$$

and with $\nu = 3/2$ we obtain $3 J_{3/2}(x) = x J_{5/2}(x) + x J_{1/2}(x)$ so that

$$J_{5/2}(x) = \sqrt{\frac{2}{\pi x}}\left(\frac{3 \sin x}{x^2} - \frac{3 \cos x}{x} - \sin x\right).$$

(b)

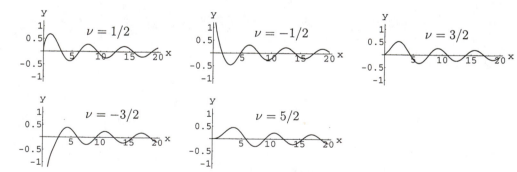

27. The general solution of Bessel's equation is

$$w(t) = c_1 J_{1/3}(t) + c_2 J_{-1/3}(t), \qquad t > 0.$$

Thus, the general solution of Airy's equation for $x > 0$ is

$$y = x^{1/2} w\left(\frac{2}{3}\alpha x^{3/2}\right) = c_1 x^{1/2} J_{1/3}\left(\frac{2}{3}\alpha x^{3/2}\right) + c_2 x^{1/2} J_{-1/3}\left(\frac{2}{3}\alpha x^{3/2}\right).$$

30. The recurrence relation can be written

$$P_{k+1}(x) = \frac{2k+1}{k+1} x P_k(x) - \frac{k}{k+1} P_{k-1}(x), \qquad k = 2,\ 3,\ 4,\ \ldots\ .$$

$k=1$: $\quad P_2(x) = \frac{3}{2}x^2 - \frac{1}{2}$

$k=2$: $\quad P_3(x) = \frac{5}{3}x\left(\frac{3}{2}x^2 - \frac{1}{2}\right) - \frac{2}{3}x = \frac{5}{2}x^3 - \frac{3}{2}x$

$k=3$: $\quad P_4(x) = \frac{7}{4}x\left(\frac{5}{2}x^3 - \frac{3}{2}x\right) - \frac{3}{4}\left(\frac{3}{2}x^2 - \frac{1}{2}\right) = \frac{35}{8}x^4 - \frac{30}{8}x^2 + \frac{3}{8}$

$k=4$: $\quad P_5(x) = \frac{9}{5}x\left(\frac{35}{8}x^4 - \frac{30}{8}x^2 + \frac{3}{8}\right) - \frac{4}{5}\left(\frac{5}{2}x^3 - \frac{3}{2}x\right) = \frac{63}{8}x^5 - \frac{35}{4}x^3 + \frac{15}{8}x$

$k=5$: $\quad P_6(x) = \frac{11}{6}x\left(\frac{63}{8}x^5 - \frac{35}{4}x^3 + \frac{15}{8}x\right) - \frac{5}{6}\left(\frac{35}{8}x^4 - \frac{30}{8}x^2 + \frac{3}{8}\right) = \frac{231}{16}x^6 - \frac{315}{16}x^4 + \frac{105}{16}x^2 - \frac{5}{16}$

$k=6$: $\quad P_7(x) = \frac{13}{7}x\left(\frac{231}{16}x^6 - \frac{315}{16}x^4 + \frac{105}{16}x^2 - \frac{5}{16}\right) - \frac{5}{6}\left(\frac{63}{8}x^5 - \frac{35}{4}x^3 + \frac{15}{8}x\right)$

$$= \frac{429}{16}x^7 - \frac{693}{16}x^5 + \frac{315}{16}x^3 - \frac{35}{16}x$$

Chapter 6 Review Exercises

3. Solving $x^2 - 2x + 10 = 0$ we obtain $x = 1 \pm \sqrt{11}$, which are singular points. Thus, the minimum radius of convergence is $|1 - \sqrt{11}| = \sqrt{11} - 1$.

6. The differential equation $(x-1)(x+3)y'' + y = 0$ has regular singular points at $x = 1$ and $x = -3$.

9. Substituting $y = \sum_{n=0}^{\infty} c_n x^n$ into the differential equation we obtain

$$(x-1)y'' + 3y = (-2c_2 + 3c_0) + \sum_{k=3}^{\infty}(k-1)(k-2)c_{k-1} - k(k-1)c_k + 3c_{k-2}]x^{k-2} = 0$$

which implies $c_2 = 3c_0/2$ and

$$c_k = \frac{(k-1)(k-2)c_{k-1} + 3c_{k-2}}{k(k-1)}, \qquad k = 3, 4, 5, \ldots\ .$$

Choosing $c_0 = 1$ and $c_1 = 0$ we find

$$c_2 = \frac{3}{2}, \qquad c_3 = \frac{1}{2}, \qquad c_4 = \frac{5}{8}$$

and so on. For $c_0 = 0$ and $c_1 = 1$ we obtain

$$c_2 = 0, \qquad c_3 = \frac{1}{2}, \qquad c_4 = \frac{1}{4}$$

and so on. Thus, two solutions are

$$y_1 = C_1 \left(1 + \frac{3}{2}x^2 + \frac{1}{2}x^3 + \frac{5}{8}x^4 + \cdots \right)$$

and

$$y_2 = C_2 \left(x + \frac{1}{2}x^3 + \frac{1}{4}x^4 + \cdots \right).$$

12. Substituting $y = \sum_{n=0}^{\infty} c_n x^n$ into the differential equation we have

$$(\cos x)y'' + y = \left(1 - \frac{1}{2}x^2 + \frac{1}{24}x^4 - \frac{1}{720}x^6 + \cdots \right)(2c_2 + 6c_3 x + 12c_4 x^2 + 20c_5 x^3 + 30c_6 x^4 + \cdots)$$

$$+ \sum_{n=0}^{\infty} c_n x^n$$

$$= \left[2c_2 + 6c_3 x + (12c_4 - c_2)x^2 + (20c_5 - 3c_3)x^3 + \left(30c_6 - 6c_4 + \frac{1}{12}c_2 \right) x^4 + \cdots \right]$$

$$+ \left[c_0 + c_1 x + c_2 x^2 + c_3 x^3 + c_4 x^4 + \cdots \right]$$

$$= (c_0 + 2c_2) + (c_1 + 6c_3)x + 12c_4 x^2 + (20c_5 - 2c_3)x^3 + \left(30c_6 - 5c_4 + \frac{1}{12}c_2 \right) x^4 + \cdots$$

$$= 0.$$

Thus

$$c_0 + 2c_2 = 0$$

$$c_1 + 6c_3 = 0$$

$$12c_4 = 0$$

$$20c_5 - 2c_3 = 0$$

$$30c_6 - 5c_4 + \frac{1}{12}c_2 = 0$$

and

$$c_2 = -\frac{1}{2}c_0$$

$$c_3 = -\frac{1}{6}c_1$$

$$c_4 = 0$$

$$c_5 = \frac{1}{10}c_3$$

$$c_6 = \frac{1}{6}c_4 - \frac{1}{360}c_2.$$

Choosing $c_0 = 1$ and $c_1 = 0$ we find

$$c_2 = -\frac{1}{2}, \quad c_3 = 0, \quad c_4 = 0, \quad c_5 = 0, \quad c_6 = \frac{1}{720}$$

and so on. For $c_0 = 0$ and $c_1 = 1$ we find

$$c_2 = 0, \quad c_3 = -\frac{1}{6}, \quad c_4 = 0, \quad c_5 = -\frac{1}{60}, \quad c_6 = 0$$

and so on. Thus, two solutions are

$$y_1 = 1 - \frac{1}{2}x^2 + \frac{1}{720}x^6 + \cdots \quad \text{and} \quad y_2 = x - \frac{1}{6}x^3 - \frac{1}{60}x^5 + \cdots.$$

15. Writing the differential equation in the form

$$y'' + \left(\frac{1 - \cos x}{x}\right)y' + xy = 0,$$

and noting that

$$\frac{1 - \cos x}{x} = \frac{x}{2} - \frac{x^3}{24} + \frac{x^5}{720} - \cdots$$

is analytic at $x = 0$, we conclude that $x = 0$ is an ordinary point of the differential equation.

18. (a) From $y = -\frac{1}{u}\frac{du}{dx}$ we obtain

$$\frac{dy}{dx} = -\frac{1}{u}\frac{d^2u}{dx^2} + \frac{1}{u^2}\left(\frac{du}{dx}\right)^2.$$

Then $dy/dx = x^2 + y^2$ becomes

$$-\frac{1}{u}\frac{d^2u}{dx^2} + \frac{1}{u^2}\left(\frac{du}{dx}\right)^2 = x^2 + \frac{1}{u^2}\left(\frac{du}{dx}\right)^2,$$

so $\dfrac{d^2u}{dx^2} + x^2 u = 0$.

(b) If $u = x^{1/2} w(\frac{1}{2}x^2)$ then

$$u' = x^{3/2}w'\left(\frac{1}{2}x^2\right) + \frac{1}{2}x^{-1/2}w\left(\frac{1}{2}x^2\right)$$

and

$$u'' = x^{5/2}w''\left(\frac{1}{2}x^2\right) + 2x^{1/2}w'\left(\frac{1}{2}x^2\right) - \frac{1}{4}x^{-3/2}w\left(\frac{1}{2}x^2\right),$$

so

$$u'' + x^2 u = x^{1/2}\left[x^2 w''\left(\frac{1}{2}x^2\right) + 2w'\left(\frac{1}{2}x^2\right) + \left(x^2 - \frac{1}{4}x^{-2}\right)w\left(\frac{1}{2}x^2\right)\right] = 0.$$

Letting $t = \frac{1}{2}x^2$ we have

$$\sqrt{2t}\left[2tw''(t) + 2w'(t) + \left(2t - \frac{1}{4 \cdot 2t}\right)w(t)\right] = 0$$

or

$$t^2 w''(t) + t w'(t) + \left(t^2 - \frac{1}{16}\right) w(t) = 0.$$

This is Bessel's equation with $\nu = 1/4$, so

$$w(t) = c_1 J_{1/4}(t) + c_2 J_{-1/4}(t).$$

(c) We have

$$y = -\frac{1}{u}\frac{du}{dx} = -\frac{1}{x^{1/2}w(t)}\frac{d}{dx}x^{1/2}w(t)$$

$$= -\frac{1}{x^{1/2}w}\left[x^{1/2}\frac{dw}{dt}\frac{dt}{dx} + \frac{1}{2}x^{-1/2}w\right]$$

$$= -\frac{1}{x^{1/2}w}\left[x^{3/2}\frac{dw}{dt} + \frac{1}{2x^{1/2}}\,w\right]$$

$$= -\frac{1}{2xw}\left[2x^2\frac{dw}{dt} + w\right] = -\frac{1}{2xw}\left[4t\frac{dw}{dt} + w\right].$$

Now

$$4t\frac{dw}{dt} + w = 4t\frac{d}{dt}[c_1 J_{1/4}(t) + c_2 J_{-1/4}(t)] + c_1 J_{1/4}(t) + c_2 J_{-1/4}(t)$$

$$= 4t\left[c_1\left(J_{-3/4}(t) - \frac{1}{4t}J_{1/4}(t)\right) + c_2\left(-\frac{1}{4t}J_{-1/4}(t) - J_{3/4}(t)\right)\right]$$

$$\qquad + c_1 J_{1/4}(t) + c_2 J_{-1/4}(t)$$

$$= 4c_1 t J_{-3/4}(t) - 4c_2 t J_{3/4}(t)$$

$$= 2c_1 x^2 J_{-3/4}\left(\frac{1}{2}x^2\right) - 2c_2 x^2 J_{3/4}\left(\frac{1}{2}x^2\right),$$

so

$$y = -\frac{2c_1 x^2 J_{-3/4}(\frac{1}{2}x^2) - 2c_2 x^2 J_{3/4}(\frac{1}{2}x^2)}{2x[c_1 J_{1/4}(\frac{1}{2}x^2) + c_2 J_{-1/4}(\frac{1}{2}x^2)]}$$

$$= x\frac{-c_1 J_{-3/4}(\frac{1}{2}x^2) + c_2 J_{3/4}(\frac{1}{2}x^2)}{c_1 J_{1/4}(\frac{1}{2}x^2) + c_2 J_{-1/4}(\frac{1}{2}x^2)}.$$

Letting $c = c_1/c_2$ we have

$$y = x\frac{J_{3/4}(\frac{1}{2}x^2) - c J_{-3/4}(\frac{1}{2}x^2)}{c J_{1/4}(\frac{1}{2}x^2) + J_{-1/4}(\frac{1}{2}x^2)}.$$

7 The Laplace Transform

Exercises 7.1

3. $\mathcal{L}\{f(t)\} = \int_0^1 te^{-st}dt + \int_1^\infty e^{-st}dt = \left(-\frac{1}{s}te^{-st} - \frac{1}{s^2}e^{-st}\right)\Big|_0^1 - \frac{1}{s}e^{-st}\Big|_1^\infty$

$= \left(-\frac{1}{s}e^{-s} - \frac{1}{s^2}e^{-s}\right) - \left(0 - \frac{1}{s^2}\right) - \frac{1}{s}(0 - e^{-s}) = \frac{1}{s^2}(1 - e^{-s}), \quad s > 0$

6. $\mathcal{L}\{f(t)\} = \int_{\pi/2}^\infty (\cos t)e^{-st}dt = \left(-\frac{s}{s^2+1}e^{-st}\cos t + \frac{1}{s^2+1}e^{-st}\sin t\right)\Big|_{\pi/2}^\infty$

$= 0 - \left(0 + \frac{1}{s^2+1}e^{-\pi s/2}\right) = -\frac{1}{s^2+1}e^{-\pi s/2}, \quad s > 0$

9. $f(t) = \begin{cases} 1 - t, & 0 < t < 1 \\ 0, & t > 0 \end{cases}$

$\mathcal{L}\{f(t)\} = \int_0^1 (1-t)e^{-st}dt = \left(-\frac{1}{s}(1-t)e^{-st} + \frac{1}{s^2}e^{-st}\right)\Big|_0^1 = \frac{1}{s^2}e^{-s} + \frac{1}{s} - \frac{1}{s^2}, \quad s > 0$

12. $\mathcal{L}\{f(t)\} = \int_0^\infty e^{-2t-5}e^{-st}dt = e^{-5}\int_0^\infty e^{-(s+2)t}dt = -\frac{e^{-5}}{s+2}e^{-(s+2)t}\Big|_0^\infty = \frac{e^{-5}}{s+2}, \quad s > -2$

15. $\mathcal{L}\{f(t)\} = \int_0^\infty e^{-t}(\sin t)e^{-st}dt = \int_0^\infty (\sin t)e^{-(s+1)t}dt$

$= \left(\frac{-(s+1)}{(s+1)^2+1}e^{-(s+1)t}\sin t - \frac{1}{(s+1)^2+1}e^{-(s+1)t}\cos t\right)\Big|_0^\infty$

$= \frac{1}{(s+1)^2+1} = \frac{1}{s^2+2s+2}, \quad s > -1$

18. $\mathcal{L}\{f(t)\} = \int_0^\infty t(\sin t)e^{-st}dt$

$= \left[\left(-\frac{t}{s^2+1} - \frac{2s}{(s^2+1)^2}\right)(\cos t)e^{-st} - \left(\frac{st}{s^2+1} + \frac{s^2-1}{(s^2+1)^2}\right)(\sin t)e^{-st}\right]_0^\infty$

$= \frac{2s}{(s^2+1)^2}, \quad s > 0$

21. $\mathcal{L}\{4t - 10\} = \frac{4}{s^2} - \frac{10}{s}$

24. $\mathcal{L}\{-4t^2 + 16t + 9\} = -4\frac{2}{s^3} + \frac{16}{s^2} + \frac{9}{s}$

27. $\mathcal{L}\{1 + e^{4t}\} = \frac{1}{s} + \frac{1}{s-4}$

30. $\mathcal{L}\{e^{2t} - 2 + e^{-2t}\} = \frac{1}{s-2} - \frac{2}{s} + \frac{1}{s+2}$

33. $\mathcal{L}\{\sinh kt\} = \dfrac{k}{s^2 - k^2}$

36. $\mathcal{L}\{e^{-t}\cosh t\} = \mathcal{L}\left\{e^{-t}\dfrac{e^t + e^{-t}}{2}\right\} = \mathcal{L}\left\{\dfrac{1}{2} + \dfrac{1}{2}e^{-2t}\right\} = \dfrac{1}{2s} + \dfrac{1}{2(s+2)}$

39. (a) Using integration by parts for $\alpha > 0$,

$$\Gamma(\alpha + 1) = \int_0^\infty t^\alpha e^{-t}\, dt = -t^\alpha e^{-t}\Big|_0^\infty + \alpha \int_0^\infty t^{\alpha-1}e^{-t}\, dt = \alpha\Gamma(\alpha).$$

(b) Let $u = st$ so that $du = s\, dt$. Then

$$\mathcal{L}\{t^\alpha\} = \int_0^\infty e^{-st}t^\alpha dt = \int_0^\infty e^{-u}\left(\dfrac{u}{s}\right)^\alpha \dfrac{1}{s}\, du = \dfrac{1}{s^{\alpha+1}}\Gamma(\alpha+1), \quad \alpha > -1.$$

Exercises 7.2

3. $\mathcal{L}^{-1}\left\{\dfrac{1}{s^2} - \dfrac{48}{s^5}\right\} = \mathcal{L}^{-1}\left\{\dfrac{1}{s^2} - \dfrac{48}{24}\cdot\dfrac{4!}{s^5}\right\} = t - 2t^4$

6. $\mathcal{L}^{-1}\left\{\dfrac{(s+2)^2}{s^3}\right\} = \mathcal{L}^{-1}\left\{\dfrac{1}{s} + 4\cdot\dfrac{1}{s^2} + 2\cdot\dfrac{2}{s^3}\right\} = 1 + 4t + 2t^2$

9. $\mathcal{L}^{-1}\left\{\dfrac{1}{4s+1}\right\} = \mathcal{L}^{-1}\left\{\dfrac{1}{4}\cdot\dfrac{1}{s+1/4}\right\} = \dfrac{1}{4}e^{-t/4}$

12. $\mathcal{L}^{-1}\left\{\dfrac{10s}{s^2+16}\right\} = 10\cos 4t$

15. $\mathcal{L}^{-1}\left\{\dfrac{2s-6}{s^2+9}\right\} = \mathcal{L}^{-1}\left\{2\cdot\dfrac{s}{s^2+9} - 2\cdot\dfrac{3}{s^2+9}\right\} = 2\cos 3t - 2\sin 3t$

18. $\mathcal{L}^{-1}\left\{\dfrac{s+1}{s^2-4s}\right\} = \mathcal{L}^{-1}\left\{-\dfrac{1}{4}\cdot\dfrac{1}{s} + \dfrac{5}{4}\cdot\dfrac{1}{s-4}\right\} = -\dfrac{1}{4} + \dfrac{5}{4}e^{4t}$

21. $\mathcal{L}^{-1}\left\{\dfrac{0.9s}{(s-0.1)(s+0.2)}\right\} = \mathcal{L}^{-1}\left\{(0.3)\cdot\dfrac{1}{s-0.1} + (0.6)\cdot\dfrac{1}{s+0.2}\right\} = 0.3e^{0.1t} + 0.6e^{-0.2t}$

24. $\mathcal{L}^{-1}\left\{\dfrac{s^2+1}{s(s-1)(s+1)(s-2)}\right\} = \mathcal{L}^{-1}\left\{\dfrac{1}{2}\cdot\dfrac{1}{s} - \dfrac{1}{s-1} - \dfrac{1}{3}\cdot\dfrac{1}{s+1} + \dfrac{5}{6}\cdot\dfrac{1}{s-2}\right\}$

$$= \dfrac{1}{2} - e^t - \dfrac{1}{3}e^{-t} + \dfrac{5}{6}e^{2t}$$

27. $\mathcal{L}^{-1}\left\{\dfrac{2s-4}{(s^2+s)(s^2+1)}\right\} = \mathcal{L}^{-1}\left\{\dfrac{2s-4}{s(s^2+1)^2}\right\} = \mathcal{L}^{-1}\left\{-\dfrac{4}{s} + \dfrac{3}{s+1} + \dfrac{s}{s^2+1} + \dfrac{3}{s^2+1}\right\}$

$$= -4 + 3e^{-t} + \cos t + 3\sin t$$

30. $\mathcal{L}^{-1}\left\{\dfrac{6s+3}{(s^2+1)(s^2+4)}\right\} = \mathcal{L}^{-1}\left\{2\cdot\dfrac{s}{s^2+1}+\dfrac{1}{s^2+1}-2\cdot\dfrac{s}{s^2+4}-\dfrac{1}{2}\cdot\dfrac{2}{s^2+4}\right\}$

$$= 2\cos t + \sin t - 2\cos 2t - \frac{1}{2}\sin 2t$$

33. The Laplace transform of the differential equation is

$$s\mathcal{L}\{y\} - y(0) + 6\mathcal{L}\{y\} = \frac{1}{s-4}.$$

Solving for $\mathcal{L}\{y\}$ we obtain

$$\mathcal{L}\{y\} = \frac{1}{(s-4)(s+6)} + \frac{2}{s+6} = \frac{1}{10}\cdot\frac{1}{s-4} + \frac{19}{10}\cdot\frac{1}{s+6}.$$

Thus

$$y = \frac{1}{10}e^{4t} + \frac{19}{10}e^{-6t}.$$

36. The Laplace transform of the differential equation is

$$s^2\mathcal{L}\{y\} - sy(0) - y'(0) - 4[s\mathcal{L}\{y\} - y(0)] = \frac{6}{s-3} - \frac{3}{s+1}.$$

Solving for $\mathcal{L}\{y\}$ we obtain

$$\mathcal{L}\{y\} = \frac{6}{(s-3)(s^2-4s)} - \frac{3}{(s+1)(s^2-4s)} + \frac{s-5}{s^2-4s}$$

$$= \frac{5}{2}\cdot\frac{1}{s} - \frac{2}{s-3} - \frac{3}{5}\cdot\frac{1}{s+1} + \frac{11}{10}\cdot\frac{1}{s-4}.$$

Thus

$$y = \frac{5}{2} - 2e^{3t} - \frac{3}{5}e^{-t} + \frac{11}{10}e^{4t}.$$

39. The Laplace transform of the differential equation is

$$2\left[s^3\mathcal{L}\{y\} - s^2(0) - sy'(0) - y''(0)\right] + 3[s^2\mathcal{L}\{y\} - sy(0) - y'(0)] - 3[s\mathcal{L}\{y\} - y(0)] - 2\mathcal{L}\{y\} = \frac{1}{s+1}.$$

Solving for $\mathcal{L}\{y\}$ we obtain

$$\mathcal{L}\{y\} = \frac{2s+3}{(s+1)(s-1)(2s+1)(s+2)} = \frac{1}{2}\frac{1}{s+1} + \frac{5}{18}\frac{1}{s-1} - \frac{8}{9}\frac{1}{s+1/2} + \frac{1}{9}\frac{1}{s+2}.$$

Thus

$$y = \frac{1}{2}e^{-t} + \frac{5}{18}e^{t} - \frac{8}{9}e^{-t/2} + \frac{1}{9}e^{-2t}.$$

Exercises 7.3

3. $\mathcal{L}\left\{t^3 e^{-2t}\right\} = \dfrac{3!}{(s+2)^4}$

6. $\mathcal{L}\left\{e^{2t}(t-1)^2\right\} = \mathcal{L}\left\{t^2 e^{2t} - 2t e^{2t} + e^{2t}\right\} = \dfrac{2}{(s-2)^3} - \dfrac{2}{(s-2)^2} + \dfrac{1}{s-2}$

9. $\mathcal{L}\left\{(1 - e^t + 3e^{-4t})\cos 5t\right\} = \mathcal{L}\left\{\cos 5t - e^t \cos 5t + 3e^{-4t}\cos 5t\right\}$

$$= \dfrac{s}{s^2+25} - \dfrac{s-1}{(s-1)^2+25} + \dfrac{3(s+4)}{(s+4)^2+25}$$

12. $\mathcal{L}^{-1}\left\{\dfrac{1}{(s-1)^4}\right\} = \mathcal{L}^{-1}\left\{\dfrac{1}{6}\dfrac{3!}{(s-1)^4}\right\} = \dfrac{1}{6}t^3 e^t$

15. $\mathcal{L}^{-1}\left\{\dfrac{s}{s^2+4s+5}\right\} = \mathcal{L}^{-1}\left\{\dfrac{(s+2)}{(s+2)^2+1^2} - 2\dfrac{1}{(s+2)^2+1^2}\right\} = e^{-2t}\cos t - 2e^{-2t}\sin t$

18. $\mathcal{L}^{-1}\left\{\dfrac{5s}{(s-2)^2}\right\} = \mathcal{L}^{-1}\left\{\dfrac{5(s-2)+10}{(s-2)^2}\right\} = \mathcal{L}^{-1}\left\{\dfrac{5}{s-2} + \dfrac{10}{(s-2)^2}\right\} = 5e^{2t} + 10t e^{2t}$

21. The Laplace transform of the differential equation is

$$s\mathcal{L}\{y\} - y(0) + 4\mathcal{L}\{y\} = \dfrac{1}{s+4}.$$

Solving for $\mathcal{L}\{y\}$ we obtain $\mathcal{L}\{y\} = \dfrac{1}{(s+4)^2} + \dfrac{2}{s+4}$. Thus

$$y = te^{-4t} + 2e^{-4t}.$$

24. The Laplace transform of the differential equation is

$$s^2\mathcal{L}\{y\} - sy(0) - y'(0) - 4\left[s\mathcal{L}\{y\} - y(0)\right] + 4\mathcal{L}\{y\} = \dfrac{6}{(s-2)^4}.$$

Solving for $\mathcal{L}\{y\}$ we obtain $\mathcal{L}\{y\} = \dfrac{1}{20}\dfrac{5!}{(s-2)^6}$. Thus, $y = \dfrac{1}{20}t^5 e^{2t}$.

27. The Laplace transform of the differential equation is

$$s^2\mathcal{L}\{y\} - sy(0) - y'(0) - 6\left[s\mathcal{L}\{y\} - y(0)\right] + 13\mathcal{L}\{y\} = 0.$$

Solving for $\mathcal{L}\{y\}$ we obtain

$$\mathcal{L}\{y\} = -\dfrac{3}{s^2-6s+13} = -\dfrac{3}{2}\dfrac{2}{(s-3)^2+2^2}.$$

Thus

$$y = -\dfrac{3}{2}e^{3t}\sin 2t.$$

30. The Laplace transform of the differential equation is

$$s^2 \mathscr{L}\{y\} - sy(0) - y'(0) - 2\left[s\mathscr{L}\{y\} - y(0)\right] + 5\mathscr{L}\{y\} = \frac{1}{s} + \frac{1}{s^2}.$$

Solving for $\mathscr{L}\{y\}$ we obtain

$$\mathscr{L}\{y\} = \frac{4s^2 + s + 1}{s^2(s^2 - 2s + 5)} = \frac{7}{25}\frac{1}{s} + \frac{1}{5}\frac{1}{s^2} + \frac{-7s/25 + 109/25}{s^2 - 2s + 5}$$

$$= \frac{7}{25}\frac{1}{s} + \frac{1}{5}\frac{1}{s^2} - \frac{7}{25}\frac{s-1}{(s-1)^2 + 2^2} + \frac{51}{25}\frac{2}{(s-1)^2 + 2^2}.$$

Thus

$$y = \frac{7}{25} + \frac{1}{5}t - \frac{7}{25}e^t \cos 2t + \frac{51}{25}e^t \sin 2t.$$

33. Recall from Section 5.1 that $mx'' = -kx - \beta x'$. Now $m = W/g = 4/32 = \frac{1}{8}$ slug, and $4 = 2k$ so that $k = 2$ lb/ft. Thus, the differential equation is $x'' + 7x' + 16x = 0$. The initial conditions are $x(0) = -3/2$ and $x'(0) = 0$. The Laplace transform of the differential equation is

$$s^2 \mathscr{L}\{x\} + \frac{3}{2}s + 7s\mathscr{L}\{x\} + \frac{21}{2} + 16\mathscr{L}\{x\} = 0.$$

Solving for $\mathscr{L}\{x\}$ we obtain

$$\mathscr{L}\{x\} = \frac{-3s/2 - 21/2}{s^2 + 7s + 16} = -\frac{3}{2}\frac{s + 7/2}{(s+7/2)^2 + (\sqrt{15}/2)^2} - \frac{7\sqrt{15}}{10}\frac{\sqrt{15}/2}{(s+7/2)^2 + (\sqrt{15}/2)^2}.$$

Thus

$$x = -\frac{3}{2}e^{-7t/2}\cos\frac{\sqrt{15}}{2}t - \frac{7\sqrt{15}}{10}e^{-7t/2}\sin\frac{\sqrt{15}}{2}t.$$

36. The differential equation is

$$R\frac{dq}{dt} + \frac{1}{C}q = E_0 e^{-kt}, \quad q(0) = 0.$$

The Laplace transform of this equation is

$$R\mathscr{L}\{q\} + \frac{1}{C}\mathscr{L}\{q\} = E_0\frac{1}{s+k}.$$

Solving for $\mathscr{L}\{q\}$ we obtain

$$\mathscr{L}\{q\} = \frac{E_0 C}{(s+k)(RC_s + 1)} = \frac{E_0/R}{(s+k)(s+1/RC)}.$$

When $1/RC \neq k$ we have by partial fractions

$$\mathscr{L}\{q\} = \frac{E_0}{R}\left(\frac{1/(1/RC - k)}{s+k} - \frac{1/(1/RC - k)}{s+1/RC}\right) = \frac{E_0}{R}\frac{1}{1/RC - k}\left(\frac{1}{s+k} - \frac{1}{s+1/RC}\right).$$

Thus

$$q(t) = \frac{E_0 C}{1 - kRC}\left(e^{-kt} - e^{-t/RC}\right).$$

When $1/RC = k$ we have

$$\mathcal{L}\{q\} = \frac{E_0}{R} \frac{1}{(s+k)^2}.$$

Thus

$$q(t) = \frac{E_0}{R} te^{-kt} = \frac{E_0}{R} te^{-t/RC}.$$

39. $\mathcal{L}\{t\,\mathcal{U}(t-2)\} = \mathcal{L}\{(t-2)\,\mathcal{U}(t-2) + 2\,\mathcal{U}(t-2)\} = \dfrac{e^{-2s}}{s^2} + \dfrac{2e^{-2s}}{s}$

42. $\mathcal{L}\left\{\sin t\,\mathcal{U}\left(t - \dfrac{\pi}{2}\right)\right\} = \mathcal{L}\left\{\cos\left(t - \dfrac{\pi}{2}\right)\mathcal{U}\left(t - \dfrac{\pi}{2}\right)\right\} = \dfrac{se^{-\pi s}}{s^2+1}$

45. $\mathcal{L}^{-1}\left\{\dfrac{e^{-\pi s}}{s^2+1}\right\} = \sin(t-\pi)\,\mathcal{U}(t-\pi)$

48. $\mathcal{L}^{-1}\left\{\dfrac{e^{-2s}}{s^2(s-1)}\right\} = \mathcal{L}^{-1}\left\{-\dfrac{e^{-2s}}{s} - \dfrac{e^{-2s}}{s^2} + \dfrac{e^{-2s}}{s-1}\right\} = -\mathcal{U}(t-2) - (t-2)\,\mathcal{U}(t-2) + e^{t-2}\,\mathcal{U}(t-2)$

51. (f)

54. (d)

57. $\mathcal{L}\{t^2\,\mathcal{U}(t-1)\} = \mathcal{L}\{[(t-1)^2 + 2t - 1]\,\mathcal{U}(t-1)\} = \mathcal{L}\{[(t-1)^2 + 2(t-1) - 1]\,\mathcal{U}(t-1)\}$

$$= \left(\frac{2}{s^3} + \frac{2}{s^2} + \frac{1}{s}\right)e^{-s}$$

60. $\mathcal{L}\{\sin t - \sin t\,\mathcal{U}(t-2\pi)\} = \mathcal{L}\{\sin t - \sin(t - 2\pi)\,\mathcal{U}(t-2\pi)\} = \dfrac{1}{s^2+1} - \dfrac{e^{-2\pi s}}{s^2+1}$

63. The Laplace transform of the differential equation is

$$s\mathcal{L}\{y\} - y(0) + \mathcal{L}\{y\} = \frac{5}{s}e^{-s}.$$

Solving for $\mathcal{L}\{y\}$ we obtain

$$\mathcal{L}\{y\} = \frac{5e^{-s}}{s(s+1)} = 5e^{-s}\left[\frac{1}{s} - \frac{1}{s+1}\right].$$

Thus

$$y = 5\,\mathcal{U}(t-1) - 5e^{-(t-1)}\,\mathcal{U}(t-1).$$

66. The Laplace transform of the differential equation is

$$s^2\mathcal{L}\{y\} - sy(0) - y'(0) + 4\mathcal{L}\{y\} = \frac{1}{s} - \frac{e^{-s}}{s}.$$

Solving for $\mathcal{L}\{y\}$ we obtain

$$\mathcal{L}\{y\} = \frac{1-s}{s(s^2+4)} - e^{-s}\frac{1}{s(s^2+4)} = \frac{1}{4}\frac{1}{s} - \frac{1}{4}\frac{s}{s^2+4} - \frac{1}{2}\frac{2}{s^2+4} - e^{-s}\left[\frac{1}{4}\frac{1}{s} - \frac{1}{4}\frac{s}{s^2+4}\right].$$

Thus

$$y = \frac{1}{4} - \frac{1}{4}\cos 2t - \frac{1}{2}\sin 2t - \left[\frac{1}{4} - \frac{1}{4}\cos 2(t-1)\right]\mathcal{U}(t-1).$$

69. The Laplace transform of the differential equation is

$$s^2\,\mathcal{L}\{y\} - sy(0) - y'(0) + \mathcal{L}\{y\} = \frac{e^{-\pi s}}{s} - \frac{e^{-2\pi s}}{s}.$$

Solving for $\mathcal{L}\{y\}$ we obtain

$$\mathcal{L}\{y\} = e^{-\pi s}\left[\frac{1}{s} - \frac{s}{s^2+1}\right] - e^{-2\pi s}\left[\frac{1}{s} - \frac{s}{s^2+1}\right] + \frac{1}{s^2+1}.$$

Thus

$$y = [1 - \cos(t-\pi)]\mathcal{U}(t-\pi) - [1 - \cos(t-2\pi)]\mathcal{U}(t-2\pi) + \sin t.$$

72. Recall from Section 5.1 that $mx'' = -kx + f(t)$. Now $m = W/g = 32/32 = 1$ slug, and $32 = 2k$ so that $k = 16$ lb/ft. Thus, the differential equation is $x'' + 16x = f(t)$. The initial conditions are $x(0) = 0$, $x'(0) = 0$. Also, since

$$f(t) = \begin{cases} \sin t, & 0 \le t < 2\pi \\ 0, & t \ge 2\pi \end{cases}$$

and $\sin t = \sin(t - 2\pi)$ we can write

$$f(t) = \sin t - \sin(t - 2\pi)\mathcal{U}(t - 2\pi).$$

The Laplace transform of the differential equation is

$$s^2\mathcal{L}\{x\} + 16\mathcal{L}\{x\} = \frac{1}{s^2+1} - \frac{1}{s^2+1}e^{-2\pi s}.$$

Solving for $\mathcal{L}\{x\}$ we obtain

$$\mathcal{L}\{x\} = \frac{1}{(s^2+16)(s^2+1)} - \frac{1}{(s^2+16)(s^2+1)}e^{-2\pi s}$$

$$= \frac{-1/15}{s^2+16} + \frac{1/15}{s^2+1} - \left[\frac{-1/15}{s^2+16} + \frac{1/15}{s^2+1}\right]e^{-2\pi s}.$$

Thus

$$x(t) = -\frac{1}{60}\sin 4t + \frac{1}{15}\sin t + \frac{1}{60}\sin 4(t-2\pi)\mathcal{U}(t-2\pi) - \frac{1}{15}\sin(t-2\pi)\mathcal{U}(t-2\pi)$$

$$= \begin{cases} -\frac{1}{60}\sin 4t + \frac{1}{15}\sin t, & 0 \le t < 2\pi \\ 0, & t \ge 2\pi. \end{cases}$$

75. (a) The differential equation is

$$\frac{di}{dt} + 10i = \sin t + \cos\left(t - \frac{3\pi}{2}\right)\mathcal{U}\left(t - \frac{3\pi}{2}\right), \quad i(0) = 0.$$

The Laplace transform of this equation is

$$s\mathcal{L}\{i\} + 10\mathcal{L}\{i\} = \frac{1}{s^2+1} + \frac{se^{-3\pi s/2}}{s^2+1}.$$

Solving for $\mathcal{L}\{i\}$ we obtain

$$\mathcal{L}\{i\} = \frac{1}{(s^2+1)(s+10)} + \frac{s}{(s^2+1)(s+10)}e^{-3\pi s/2}$$

$$= \frac{1}{101}\left(\frac{1}{s+10} - \frac{s}{s^2+1} + \frac{10}{s^2+1}\right) + \frac{1}{101}\left(\frac{-10}{s+10} + \frac{10s}{s^2+1} + \frac{1}{s^2+1}\right)e^{-3\pi s/2}.$$

Thus

$$i(t) = \frac{1}{101}\left(e^{-10t} - \cos t + 10\sin t\right)$$

$$+ \frac{1}{101}\left[-10e^{-10(t-3\pi/2)} + 10\cos\left(t - \frac{3\pi}{2}\right) + \sin\left(t - \frac{3\pi}{2}\right)\right]\mathcal{U}\left(t - \frac{3\pi}{2}\right).$$

(b)

The maximum value of $i(t)$ is approximately 0.1 at $t = 1.7$, the minimum is approximately -0.1 at 4.7.

78. The differential equation is

$$EI\frac{d^4y}{dx^4} = w_0[\mathcal{U}(x - L/3) - \mathcal{U}(x - 2L/3)].$$

Taking the Laplace transform of both sides and using $y(0) = y'(0) = 0$ we obtain

$$s^4\mathcal{L}\{y\} - sy''(0) - y'''(0) = \frac{w_0}{EI}\frac{1}{s}\left(e^{-Ls/3} - e^{-2Ls/3}\right).$$

Letting $y''(0) = c_1$ and $y'''(0) = c_2$ we have

$$\mathcal{L}\{y\} = \frac{c_1}{s^3} + \frac{c_2}{s^4} + \frac{w_0}{EI}\frac{1}{s^5}\left(e^{-Ls/3} - e^{-2Ls/3}\right)$$

so that

$$y(x) = \frac{1}{2}c_1x^2 + \frac{1}{6}c_2x^3 + \frac{1}{24}\frac{w_0}{EI}\left[\left(x - \frac{L}{3}\right)^4\mathcal{U}\left(x - \frac{L}{3}\right) - \left(x - \frac{2L}{3}\right)^4\mathcal{U}\left(x - \frac{2L}{3}\right)\right].$$

To find c_1 and c_2 we compute

$$y''(x) = c_1 + c_2x + \frac{1}{2}\frac{w_0}{EI}\left[\left(x - \frac{L}{3}\right)^2\mathcal{U}\left(x - \frac{L}{3}\right) - \left(x - \frac{2L}{3}\right)^2\mathcal{U}\left(x - \frac{2L}{3}\right)\right].$$

and

$$y'''(x) = c_2 + \frac{w_0}{EI}\left[\left(x - \frac{L}{3}\right)\mathcal{U}\left(x - \frac{L}{3}\right) - \left(x - \frac{2L}{3}\right)\mathcal{U}\left(x - \frac{2L}{3}\right)\right].$$

Then $y''(L) = y'''(L) = 0$ yields the system

$$c_1 + c_2 L + \frac{1}{2}\frac{w_0}{EI}\left[\left(\frac{2L}{3}\right)^2 - \left(\frac{L}{3}\right)^2\right] = c_1 + c_2 L + \frac{1}{6}\frac{w_0 L^2}{EI} = 0$$

$$c_2 + \frac{w_0}{EI}\left[\frac{2L}{3} - \frac{L}{3}\right] = c_2 + \frac{1}{3}\frac{w_0 L}{EI} = 0.$$

Solving for c_1 and c_2 we obtain $c_1 = \frac{1}{6}w_0 L^2/EI$ and $c_2 = -\frac{1}{3}w_0 L/EI$. Thus

$$y(x) = \frac{w_0}{EI}\left(\frac{1}{12}L^2 x^2 - \frac{1}{18}Lx^3 + \frac{1}{24}\left[\left(x - \frac{L}{3}\right)^4 \mathcal{U}\left(x - \frac{L}{3}\right) - \left(x - \frac{2L}{3}\right)^4 \mathcal{U}\left(x - \frac{2L}{3}\right)\right]\right).$$

Exercises 7.4

3. $\mathcal{L}\{t^2 \sinh t\} = \dfrac{d^2}{ds^2}\left(\dfrac{1}{s^2 - 1}\right) = \dfrac{6s^2 + 2}{(s^2 - 1)^3}$

6. $\mathcal{L}\{te^{-3t}\cos 3t\} = -\dfrac{d}{ds}\left(\dfrac{s + 3}{(s+3)^2 + 9}\right) = \dfrac{(s+3)^2 - 9}{[(s+3)^2 + 9]^2}$

9. $\mathcal{L}\{e^{-t} * e^t \cos t\} = \dfrac{s - 1}{(s+1)\left[(s-1)^2 + 1\right]}$

12. $\mathcal{L}\left\{\displaystyle\int_0^t \cos\tau\, d\tau\right\} = \dfrac{1}{s}\mathcal{L}\{\cos t\} = \dfrac{s}{s(s^2 + 1)} = \dfrac{1}{s^2 + 1}$

15. $\mathcal{L}\left\{\displaystyle\int_0^t \tau e^{t-\tau}\, d\tau\right\} = \mathcal{L}\{t\}\mathcal{L}\{e^t\} = \dfrac{1}{s^2(s - 1)}$

18. $\mathcal{L}\left\{t\displaystyle\int_0^t \tau e^{-\tau} d\tau\right\} = -\dfrac{d}{ds}\mathcal{L}\left\{t\displaystyle\int_0^t \tau e^{-\tau} d\tau\right\} = -\dfrac{d}{ds}\left(\dfrac{1}{s}\dfrac{1}{(s+1)^2}\right) = \dfrac{3s + 1}{s^2(s+1)^3}$

21. $\mathcal{L}\{f(t)\} = \dfrac{1}{1 - e^{-2as}}\left[\displaystyle\int_0^a e^{-st}dt - \int_a^{2a} e^{-st}dt\right] = \dfrac{(1 - e^{-as})^2}{s(1 - e^{-2as})} = \dfrac{1 - e^{-as}}{s(1 + e^{-as})}$

24. $\mathcal{L}\{f(t)\} = \dfrac{1}{1 - e^{-2s}}\left[\displaystyle\int_0^1 te^{-st}dt + \int_1^2 (2 - t)e^{-st}dt\right] = \dfrac{1 - e^{-s}}{s^2(1 - e^{-2s})}$

27. The Laplace transform of the differential equation is

$$s\mathcal{L}\{y\} + \mathcal{L}\{y\} = \dfrac{2s}{(s^2 + 1)^2}.$$

Solving for $\mathscr{L}\{y\}$ we obtain

$$\mathscr{L}\{y\} = \frac{2s}{(s+1)(s^2+1)^2} = -\frac{1}{2}\frac{1}{s+1} - \frac{1}{2}\frac{1}{s^2+1} + \frac{1}{2}\frac{s}{s^2+1} + \frac{1}{(s^2+1)^2} + \frac{s}{(s^2+1)^2}.$$

Thus

$$y(t) = -\frac{1}{2}e^{-t} - \frac{1}{2}\sin t + \frac{1}{2}\cos t + \frac{1}{2}(\sin t - t\cos t) + \frac{1}{2}t\sin t$$

$$= -\frac{1}{2}e^{-t} + \frac{1}{2}\cos t - \frac{1}{2}t\cos t + \frac{1}{2}t\sin t.$$

30. The Laplace transform of the differential equation is

$$s^2\mathscr{L}\{y\} - sy(0) - y'(0) + \mathscr{L}\{y\} = \frac{1}{s^2+1}.$$

Solving for $\mathscr{L}\{y\}$ we obtain

$$\mathscr{L}\{y\} = \frac{s^3 - s^2 + s}{(s^2+1)^2} = \frac{s}{s^2+1} - \frac{1}{s^2+1} + \frac{1}{(s^2+1)^2}.$$

Thus

$$y = \cos t - \frac{1}{2}\sin t - \frac{1}{2}t\cos t.$$

33. The Laplace transform of the differential equation is

$$s^2\mathscr{L}\{y\} + \mathscr{L}\{y\} = \frac{1}{(s^2+1)} + \frac{2s}{(s^2+1)^2}.$$

Thus

$$\mathscr{L}\{y\} = \frac{1}{(s^2+1)^2} + \frac{2s}{(s^2+1)^3}$$

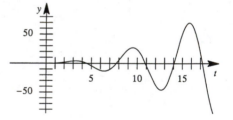

In Problem 20, it is shown that

$$\mathscr{L}^{-1}\left\{\frac{8k^3s}{(s^2+k^2)^3}\right\} = t\sin kt - kt^2\cos kt.$$

Applying this result we find that

$$y = \frac{1}{2}(\sin t - t\cos t) + \frac{1}{4}(t\sin t - t^2\cos t).$$

36. The Laplace transform of the given equation is

$$\mathscr{L}\{f\} = \mathscr{L}\{2t\} - 4\mathscr{L}\{\sin t\}\mathscr{L}\{f\}.$$

Solving for $\mathscr{L}\{f\}$ we obtain

$$\mathscr{L}\{f\} = \frac{2s^2+2}{s^2(s^2+5)} = \frac{2}{5}\frac{1}{s^2} + \frac{8}{5\sqrt{5}}\frac{\sqrt{5}}{s^2+5}.$$

Thus

$$f(t) = \frac{2}{5}t + \frac{8}{5\sqrt{5}}\sin\sqrt{5}\,t.$$

39. The Laplace transform of the given equation is

$$\mathcal{L}\{f\} + \mathcal{L}\{1\}\,\mathcal{L}\{f\} = \mathcal{L}\{1\}.$$

Solving for $\mathcal{L}\{f\}$ we obtain $\mathcal{L}\{f\} = \dfrac{1}{s+1}$. Thus, $f(t) = e^{-t}$.

42. The Laplace transform of the given equation is

$$\mathcal{L}\{t\} - 2\,\mathcal{L}\{f\} = \mathcal{L}\left\{e^t - e^{-t}\right\}\mathcal{L}\{f\}.$$

Solving for $\mathcal{L}\{f\}$ we obtain

$$\mathcal{L}\{f\} = \frac{s^2-1}{2s^4} = \frac{1}{2}\frac{1}{s^2} - \frac{1}{12}\frac{3!}{s^4}.$$

Thus

$$f(t) = \frac{1}{2}t - \frac{1}{12}t^3.$$

45. The differential equation is

$$0.1\frac{di}{dt} + 3i + \frac{1}{0.05}\int_0^t i(\tau)d\tau = 100[\mathcal{U}(t-1) - \mathcal{U}(t-2)]$$

or

$$\frac{di}{st} + 30i + 200\int_0^t i(\tau)d\tau = 1000[\,\mathcal{U}(t-1) - \mathcal{U}(t-2)],$$

where $i(0) = 0$. The Laplace transform of the differential equation is

$$s\,\mathcal{L}\{i\} - y(0) + 30\,\mathcal{L}\{i\} + \frac{200}{s}\mathcal{L}\{i\} = \frac{1000}{s}(e^{-s} - e^{-2s}).$$

Solving for $\mathcal{L}\{i\}$ we obtain

$$\mathcal{L}\{i\} = \frac{1000e^{-s} - 1000e^{-2s}}{s^2 + 30s + 200} = \left(\frac{100}{s+10} - \frac{100}{s+20}\right)(e^{-s} - e^{-2s}).$$

Thus

$$i(t) = 100(e^{-10(t-1)} - e^{-20(t-1)})\,\mathcal{U}(t-1) - 100(e^{-10(t-2)} - e^{-20(t-2)})\,\mathcal{U}(t-2).$$

48. The differential equation is

$$\frac{di}{dt} + i = E(t),$$

where $i(0) = 0$. The Laplace transform of this equation is

$$s\mathcal{L}\{i\} + \mathcal{L}\{i\} = \mathcal{L}\{E(t)\}.$$

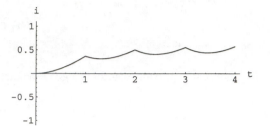

From Problem 23 we have

$$\mathcal{L}\{E(t)\} = \frac{1}{s}\left(\frac{1}{s} - \frac{1}{e^s - 1}\right) = \frac{1}{s^2} - \frac{1}{s}\frac{1}{e^s - 1}.$$

Thus

$$(s+1)\mathcal{L}\{i\} = \frac{1}{s^2} - \frac{1}{s}\frac{1}{e^s - 1}$$

and

$$\mathcal{L}\{i\} = \frac{1}{s^2(s+1)} - \frac{1}{s(s+1)}\frac{1}{e^s - 1}$$

$$= \left(\frac{1}{s^2} - \frac{1}{s} + \frac{1}{s+1}\right) - \left(\frac{1}{s} - \frac{1}{s+1}\right)\frac{1}{e^s - 1}$$

$$= \left(\frac{1}{s^2} - \frac{1}{s} + \frac{1}{s+1}\right) - \left(\frac{1}{s} - \frac{1}{s+1}\right)\left(e^{-s} + e^{-2s} + e^{-3s} + e^{-4s} - \cdots\right).$$

Therefore

$$i(t) = (t - 1 + e^{-t}) - (1 - e^{-(t-1)})\,\mathcal{U}(t-1) - (1 - e^{-(t-2)})\,\mathcal{U}(t-2)$$

$$- (1 - e^{-(t-3)})\,\mathcal{U}(t-3) - (1 - e^{-(t-4)})\,\mathcal{U}(t-4) - \cdots$$

$$= (t - 1 + e^{-t}) - \sum_{n=1}^{\infty}(1 - e^{-(t-n)})\,\mathcal{U}(t-n).$$

Exercises 7.5

3. The Laplace transform of the differential equation yields

$$\mathcal{L}\{y\} = \frac{1}{s^2 + 1}\left(1 + e^{-2\pi s}\right).$$

so that

$$y = \sin t + \sin t\,\mathcal{U}(t - 2\pi).$$

6. The Laplace transform of the differential equation yields

$$\mathcal{L}\{y\} = \frac{s}{s^2 + 1} + \frac{1}{s^2 + 1}(e^{-2\pi s} + e^{-4\pi s})$$

so that

$$y = \cos t + \sin t [\mathcal{U}(t - 2\pi) + \mathcal{U}(t - 4\pi)].$$

9. The Laplace transform of the differential equation yields

$$\mathcal{L}\{y\} = \frac{1}{(s+2)^2 + 1} e^{-2\pi s}$$

so that

$$y = e^{-2(t-2\pi)} \sin t \, \mathcal{U}(t - 2\pi).$$

12. The Laplace transform of the differential equation yields

$$\mathcal{L}\{y\} = \frac{1}{(s-1)^2(s-6)} + \frac{e^{-2s} + e^{-4s}}{(s-1)(s-6)}$$

$$= -\frac{1}{25}\frac{1}{s-1} - \frac{1}{5}\frac{1}{(s-1)^2} + \frac{1}{25}\frac{1}{s-6} + \left[-\frac{1}{5}\frac{1}{s-1} + \frac{1}{5}\frac{1}{s-6}\right]\left(e^{-2s} + e^{-4s}\right)$$

so that

$$y = -\frac{1}{25}e^t - \frac{1}{5}te^t + \frac{1}{25}e^{6t} + \left[-\frac{1}{5}e^{t-2} + \frac{1}{5}e^{6(t-2)}\right]\mathcal{U}(t-2)$$

$$+ \left[-\frac{1}{5}e^{t-4} + \frac{1}{5}e^{6(t-4)}\right]\mathcal{U}(t-4).$$

--- **Exercises 7.6** ---

3. Taking the Laplace transform of the system gives

$$s\mathcal{L}\{x\} + 1 = \mathcal{L}\{x\} - 2\mathcal{L}\{y\}$$

$$s\mathcal{L}\{y\} - 2 = 5\mathcal{L}\{x\} - \mathcal{L}\{y\}$$

so that

$$\mathcal{L}\{x\} = \frac{-s-5}{s^2+9} = -\frac{s}{s^2+9} - \frac{5}{3}\frac{3}{s^2+9}$$

and

$$x = -\cos 3t - \frac{5}{3}\sin 3t.$$

Then

$$y = \frac{1}{2}x - \frac{1}{2}x' = 2\cos 3t - \frac{7}{3}\sin 3t.$$

6. Taking the Laplace transform of the system gives

$$(s+1)\mathcal{L}\{x\} - (s-1)\mathcal{L}\{y\} = -1$$

$$s\mathcal{L}\{x\} + (s+2)\mathcal{L}\{y\} = 1$$

so that

$$\mathcal{L}\{y\} = \frac{s + 1/2}{s^2 + s + 1} = \frac{s + 1/2}{(s + 1/2)^2 + (\sqrt{3}/2)^2}$$

and

$$\mathcal{L}\{x\} = \frac{-3/2}{s^2 + s + 1} = \frac{-3/2}{(s + 1/2)^2 + (\sqrt{3}/2)^2} \,.$$

Then

$$y = e^{-t/2} \cos\frac{\sqrt{3}}{2}t \quad \text{and} \quad x = e^{-t/2} \sin\frac{\sqrt{3}}{2}t.$$

9. Adding the equations and then subtracting them gives

$$\frac{d^2x}{dt^2} = \frac{1}{2}t^2 + 2t$$

$$\frac{d^2y}{dt^2} = \frac{1}{2}t^2 - 2t.$$

Taking the Laplace transform of the system gives

$$\mathcal{L}\{x\} = 8\frac{1}{s} + \frac{1}{24}\frac{4!}{s^5} + \frac{1}{3}\frac{3!}{s^4}$$

and

$$\mathcal{L}\{y\} = \frac{1}{24}\frac{4!}{s^5} - \frac{1}{3}\frac{3!}{s^4}$$

so that

$$x = 8 + \frac{1}{24}t^4 + \frac{1}{3}t^3 \quad \text{and} \quad y = \frac{1}{24}t^4 - \frac{1}{3}t^3.$$

12. Taking the Laplace transform of the system gives

$$(s - 4)\,\mathcal{L}\{x\} + 2\mathcal{L}\{y\} = \frac{2e^{-s}}{s}$$

$$-3\,\mathcal{L}\{x\} + (s + 1)\,\mathcal{L}\{y\} = \frac{1}{2} + \frac{e^{-s}}{s}$$

so that

$$\mathcal{L}\{x\} = \frac{-1/2}{(s - 1)(s - 2)} + e^{-s}\frac{1}{(s - 1)(s - 2)}$$

$$= \left[\frac{1}{2}\frac{1}{s - 1} - \frac{1}{2}\frac{1}{s - 2}\right] + e^{-s}\left[-\frac{1}{s - 1} + \frac{1}{s - 2}\right]$$

and

$$\mathcal{L}\{y\} = \frac{e^{-s}}{s} + \frac{s/4 - 1}{(s - 1)(s - 2)} + e^{-s}\frac{-s/2 + 2}{(s - 1)(s - 2)}$$

$$= \frac{3}{4}\frac{1}{s - 1} - \frac{1}{2}\frac{1}{s - 2} + e^{-s}\left[\frac{1}{s} - \frac{3}{2}\frac{1}{s - 1} + \frac{1}{s - 2}\right].$$

Then

$$x = \frac{1}{2}e^t - \frac{1}{2}e^{2t} + \left[-e^{t-1} + e^{2(t-1)}\right]\mathcal{U}(t-1)$$

and

$$y = \frac{3}{4}e^t - \frac{1}{2}e^{2t} + \left[1 - \frac{3}{2}e^{t-1} + e^{2(t-1)}\right]\mathcal{U}(t-1).$$

15. (a) By Kirchoff's first law we have $i_1 = i_2 + i_3$. By Kirchoff's second law, on each loop we have
$E(t) = Ri_1 + L_1 i_2'$ and $E(t) = Ri_1 + L_2 i_3'$ or $L_1 i_2' + Ri_2 + Ri_3 = E(t)$ and $L_2 i_3' + Ri_2 + Ri_3 = E(t)$.

(b) Taking the Laplace transform of the system

$$0.01 i_2' + 5 i_2 + 5 i_3 = 100$$

$$0.0125 i_3' + 5 i_2 + 5 i_3 = 100$$

gives

$$(s + 500)\,\mathcal{L}\{i_2\} + 500\mathcal{L}\{i_3\} = \frac{10{,}000}{s}$$

$$400\mathcal{L}\{i_2\} + (s + 400)\,\mathcal{L}\{i_3\} = \frac{8{,}000}{s}$$

so that

$$\mathcal{L}\{i_3\} = \frac{8{,}000}{s^2 + 900s} = \frac{80}{9}\frac{1}{s} - \frac{80}{9}\frac{1}{s + 900}.$$

Then

$$i_3 = \frac{80}{9} - \frac{80}{9}e^{-900t} \quad \text{and} \quad i_2 = 20 - 0.0025 i_3' - i_3 = \frac{100}{9} - \frac{100}{9}e^{-900t}.$$

(c) $i_1 = i_2 + i_3 = 20 - 20e^{-900t}$

18. Taking the Laplace transform of the system

$$0.5 i_1' + 50 i_2 = 60$$

$$0.005 i_2' + i_2 - i_1 = 0$$

gives

$$s\,\mathcal{L}\{i_1\} + 100\,\mathcal{L}\{i_2\} = \frac{120}{s}$$

$$-200\,\mathcal{L}\{i_1\} + (s + 200)\,\mathcal{L}\{i_2\} = 0$$

so that

$$\mathcal{L}\{i_2\} = \frac{24{,}000}{s(s^2 + 200s + 20{,}000)} = \frac{6}{5}\frac{1}{s} - \frac{6}{5}\frac{s + 100}{(s + 100)^2 + 100^2} - \frac{6}{5}\frac{100}{(s + 100)^2 + 100^2}.$$

Then

$$i_2 = \frac{6}{5} - \frac{6}{5}e^{-100t}\cos 100t - \frac{6}{5}e^{-100t}\sin 100t$$

and

97

$$i_1 = 0.005i_2' + i_2 = \frac{6}{5} - \frac{6}{5}e^{-100t}\cos 100t.$$

Chapter 7 Review Exercises

3. False; consider $f(t) = t^{-1/2}$.

6. False; consider $f(t) = 1$ and $g(t) = 1$.

9. $\mathscr{L}\{\sin 2t\} = \dfrac{2}{s^2 + 4}$

12. $\mathscr{L}\{\sin 2t\, \mathscr{U}(t - \pi)\} = \mathscr{L}\{\sin 2(t - \pi)\mathscr{U}(t - \pi)\} = \dfrac{2}{s^2 + 4}e^{-\pi s}$

15. $\mathscr{L}^{-1}\left\{\dfrac{1}{(s-5)^3}\right\} = \mathscr{L}^{-1}\left\{\dfrac{1}{2}\dfrac{2}{(s-5)^3}\right\} = \dfrac{1}{2}t^2 e^{5t}$

18. $\mathscr{L}^{-1}\left\{\dfrac{1}{s^2}e^{-5s}\right\} = (t-5)\mathscr{U}(t-5)$

21. $\mathscr{L}\{e^{-5t}\}$ exists for $s > -5$.

24. $\mathscr{L}\left\{\displaystyle\int_0^t e^{a\tau}f(\tau)\,d\tau\right\} = \dfrac{1}{s}\mathscr{L}\{e^{at}f(t)\} = \dfrac{F(s-a)}{s}$, whereas

$$\mathscr{L}\left\{e^{at}\int_0^t f(\tau)\,d\tau\right\} = \mathscr{L}\left\{\int_0^t f(\tau)\,d\tau\right\}\Big|_{s\to s-a} = \dfrac{F(s)}{s}\Big|_{s\to s-a} = \dfrac{F(s-a)}{s-a}.$$

27. $f(t - t_0)\mathscr{U}(t - t_0)$

30. $f(t) = \sin t\, \mathscr{U}(t - \pi) - \sin t\, \mathscr{U}(t - 3\pi) = -\sin(t - \pi)\mathscr{U}(t - \pi) + \sin(t - 3\pi)\mathscr{U}(t - 3\pi)$

$$\mathscr{L}\{f(t)\} = -\dfrac{1}{s^2 + 1}e^{-\pi s} + \dfrac{1}{s^2 + 1}e^{-3\pi s}$$

$$\mathscr{L}\{e^t f(t)\} = -\dfrac{1}{(s-1)^2 + 1}e^{-\pi(s-1)} + \dfrac{1}{(s-1)^2 + 1}e^{-3\pi(s-1)}$$

33. Taking the Laplace transform of the differential equation we obtain

$$\mathscr{L}\{y\} = \dfrac{5}{(s-1)^2} + \dfrac{1}{2}\dfrac{2}{(s-1)^3}$$

so that

$$y = 5te^t + \dfrac{1}{2}t^2 e^t.$$

36. Taking the Laplace transform of the differential equation we obtain

$$\mathcal{L}\{y\} = \frac{s^3 + 2}{s^3(s-5)} - \frac{2 + 2s + s^2}{s^3(s-5)}e^{-s}$$

$$= -\frac{2}{125}\frac{1}{s} - \frac{2}{25}\frac{1}{s^2} - \frac{1}{5}\frac{2}{s^3} + \frac{127}{125}\frac{1}{s-5} - \left[-\frac{37}{125}\frac{1}{s} - \frac{12}{25}\frac{1}{s^2} - \frac{1}{5}\frac{2}{s^3} + \frac{37}{125}\frac{1}{s-5}\right]e^{-s}$$

so that

$$y = -\frac{2}{125} - \frac{2}{25}t - \frac{1}{5}t^2 + \frac{127}{125}e^{5t} - \left[-\frac{37}{125} - \frac{12}{25}(t-1) - \frac{1}{5}(t-1)^2 + \frac{37}{125}e^{5(t-1)}\right]\mathcal{U}(t-1).$$

39. Taking the Laplace transform of the system gives

$$s\mathcal{L}\{x\} + \mathcal{L}\{y\} = \frac{1}{s^2} + 1$$

$$4\mathcal{L}\{x\} + s\mathcal{L}\{y\} = 2$$

so that

$$\mathcal{L}\{x\} = \frac{s^2 - 2s + 1}{s(s-2)(s+2)} = -\frac{1}{4}\frac{1}{s} + \frac{1}{8}\frac{1}{s-2} + \frac{9}{8}\frac{1}{s+2}.$$

Then

$$x = -\frac{1}{4} + \frac{1}{8}e^{2t} + \frac{9}{8}e^{-2t} \quad \text{and} \quad y = -x' + t = \frac{9}{4}e^{-2t} - \frac{1}{4}e^{2t} + t.$$

42. The differential equation is

$$\frac{1}{2}\frac{d^2q}{dt^2} + 10\frac{dq}{dt} + 100q = 10 - 10\mathcal{U}(t-5).$$

Taking the Laplace transform we obtain

$$\mathcal{L}\{q\} = \frac{20}{2(s^2 + 20s + 200)}\left(1 - e^{-5s}\right)$$

$$= \left[\frac{1}{10}\frac{1}{s} - \frac{1}{10}\frac{s+10}{(s+10)^2 + 10^2} - \frac{1}{10}\frac{10}{(s+10)^2 + 10^2}\right]\left(1 - e^{-5s}\right)$$

so that

$$q(t) = \frac{1}{10} - \frac{1}{10}e^{-10t}\cos 10t - \frac{1}{10}e^{-10t}\sin 10t$$

$$- \left[\frac{1}{10} - \frac{1}{10}e^{-10(t-5)}\cos 10(t-5) - \frac{1}{10}e^{-10(t-5)}\sin 10(t-5)\right]\mathcal{U}(t-5).$$

8 Systems of Linear First-Order Differential Equations

──────── **Exercises 8.1** ────────

3. Let $\mathbf{X} = \begin{pmatrix} x \\ y \\ z \end{pmatrix}$. Then

$$\mathbf{X}' = \begin{pmatrix} -3 & 4 & -9 \\ 6 & -1 & 0 \\ 10 & 4 & 3 \end{pmatrix} \mathbf{X}.$$

6. Let $\mathbf{X} = \begin{pmatrix} x \\ y \\ z \end{pmatrix}$. Then

$$\mathbf{X}' = \begin{pmatrix} -3 & 4 & 0 \\ 5 & 9 & 0 \\ 0 & 1 & 6 \end{pmatrix} \mathbf{X} + \begin{pmatrix} e^{-t} \sin 2t \\ 4e^{-t} \cos 2t \\ -e^{-t} \end{pmatrix}$$

9. $\dfrac{dx}{dt} = x - y + 2z + e^{-t} - 3t; \quad \dfrac{dy}{dt} = 3x - 4y + z + 2e^{-t} + t; \quad \dfrac{dz}{dt} = -2x + 5y + 6z + 2e^{-t} - t$

12. Since

$$\mathbf{X}' = \begin{pmatrix} 5\cos t - 5\sin t \\ 2\cos t - 4\sin t \end{pmatrix} e^t \quad \text{and} \quad \begin{pmatrix} -2 & 5 \\ -2 & 4 \end{pmatrix} \mathbf{X} = \begin{pmatrix} 5\cos t - 5\sin t \\ 2\cos t - 4\sin t \end{pmatrix} e^t$$

we see that

$$\mathbf{X}' = \begin{pmatrix} -2 & 5 \\ -2 & 4 \end{pmatrix} \mathbf{X}.$$

15. Since

$$\mathbf{X}' = \begin{pmatrix} 0 \\ 0 \\ 0 \end{pmatrix} \quad \text{and} \quad \begin{pmatrix} 1 & 2 & 1 \\ 6 & -1 & 0 \\ -1 & -2 & -1 \end{pmatrix} \mathbf{X} = \begin{pmatrix} 0 \\ 0 \\ 0 \end{pmatrix}$$

we see that

$$\mathbf{X}' = \begin{pmatrix} 1 & 2 & 1 \\ 6 & -1 & 0 \\ -1 & -2 & -1 \end{pmatrix} \mathbf{X}.$$

18. Yes, since $W(\mathbf{X}_1, \mathbf{X}_2) = 8e^{2t} \neq 0$ and $\mathbf{X}_1$ and $\mathbf{X}_2$ are linearly independent on $-\infty < t < \infty$.

21. Since

$$\mathbf{X}_p' = \begin{pmatrix} 2 \\ -1 \end{pmatrix} \quad \text{and} \quad \begin{pmatrix} 1 & 4 \\ 3 & 2 \end{pmatrix}\mathbf{X}_p + \begin{pmatrix} 2 \\ -4 \end{pmatrix}t + \begin{pmatrix} -7 \\ -18 \end{pmatrix} = \begin{pmatrix} 2 \\ -1 \end{pmatrix}$$

we see that

$$\mathbf{X}_p' = \begin{pmatrix} 1 & 4 \\ 3 & 2 \end{pmatrix}\mathbf{X}_p + \begin{pmatrix} 2 \\ -4 \end{pmatrix}t + \begin{pmatrix} -7 \\ -18 \end{pmatrix}.$$

24. Since

$$\mathbf{X}_p' = \begin{pmatrix} 3\cos 3t \\ 0 \\ -3\sin 3t \end{pmatrix} \quad \text{and} \quad \begin{pmatrix} 1 & 2 & 3 \\ -4 & 2 & 0 \\ -6 & 1 & 0 \end{pmatrix}\mathbf{X}_p + \begin{pmatrix} -1 \\ 4 \\ 3 \end{pmatrix}\sin 3t = \begin{pmatrix} 3\cos 3t \\ 0 \\ -3\sin 3t \end{pmatrix}$$

we see that

$$\mathbf{X}_p' = \begin{pmatrix} 1 & 2 & 3 \\ -4 & 2 & 0 \\ -6 & 1 & 0 \end{pmatrix}\mathbf{X}_p + \begin{pmatrix} -1 \\ 4 \\ 3 \end{pmatrix}\sin 3t.$$

Exercises 8.2

3. The system is

$$\mathbf{X}' = \begin{pmatrix} -4 & 2 \\ -5/2 & 2 \end{pmatrix}\mathbf{X}$$

and $\det(\mathbf{A} - \lambda\mathbf{I}) = (\lambda - 1)(\lambda + 3) = 0$. For $\lambda_1 = 1$ we obtain

$$\begin{pmatrix} -5 & 2 & | & 0 \\ -5/2 & 1 & | & 0 \end{pmatrix} \implies \begin{pmatrix} -5 & 2 & | & 0 \\ 0 & 0 & | & 0 \end{pmatrix} \quad \text{so that} \quad \mathbf{K}_1 = \begin{pmatrix} 2 \\ 5 \end{pmatrix}.$$

For $\lambda_2 = -3$ we obtain

$$\begin{pmatrix} -1 & 2 & | & 0 \\ -5/2 & 5 & | & 0 \end{pmatrix} \implies \begin{pmatrix} -1 & 2 & | & 0 \\ 0 & 0 & | & 0 \end{pmatrix} \quad \text{so that} \quad \mathbf{K}_2 = \begin{pmatrix} 2 \\ 1 \end{pmatrix}.$$

Then

$$\mathbf{X} = c_1 \begin{pmatrix} 2 \\ 5 \end{pmatrix} e^t + c_2 \begin{pmatrix} 2 \\ 1 \end{pmatrix} e^{-3t}.$$

6. The system is

$$\mathbf{X}' = \begin{pmatrix} -6 & 2 \\ -3 & 1 \end{pmatrix}\mathbf{X}$$

and $\det(\mathbf{A} - \lambda\mathbf{I}) = \lambda(\lambda + 5) = 0$. For $\lambda_1 = 0$ we obtain

$$\begin{pmatrix} -6 & 2 & | & 0 \\ -3 & 1 & | & 0 \end{pmatrix} \implies \begin{pmatrix} 1 & -1/3 & | & 0 \\ 0 & 0 & | & 0 \end{pmatrix} \quad \text{so that} \quad \mathbf{K}_1 = \begin{pmatrix} 1 \\ 3 \end{pmatrix}.$$

For $\lambda_2 = -5$ we obtain

$$\begin{pmatrix} -1 & 2 & | & 0 \\ -3 & 6 & | & 0 \end{pmatrix} \implies \begin{pmatrix} 1 & -2 & | & 0 \\ 0 & 0 & | & 0 \end{pmatrix} \quad \text{so that} \quad K_2 = \begin{pmatrix} 2 \\ 1 \end{pmatrix}.$$

Then

$$X = c_1 \begin{pmatrix} 1 \\ 3 \end{pmatrix} + c_2 \begin{pmatrix} 2 \\ 1 \end{pmatrix} e^{-5t}.$$

9. We have $\det(A - \lambda I) = -(\lambda+1)(\lambda-3)(\lambda+2) = 0$. For $\lambda_1 = -1$, $\lambda_2 = 3$, and $\lambda_3 = -2$ we obtain

$$K_1 = \begin{pmatrix} -1 \\ 0 \\ 1 \end{pmatrix}, \quad K_2 = \begin{pmatrix} 1 \\ 4 \\ 3 \end{pmatrix}, \quad \text{and} \quad K_3 = \begin{pmatrix} 1 \\ -1 \\ 3 \end{pmatrix},$$

so that

$$X = c_1 \begin{pmatrix} -1 \\ 0 \\ 1 \end{pmatrix} e^{-t} + c_2 \begin{pmatrix} 1 \\ 4 \\ 3 \end{pmatrix} e^{3t} + c_3 \begin{pmatrix} 1 \\ -1 \\ 3 \end{pmatrix} e^{-2t}.$$

12. We have $\det(A - \lambda I) = (\lambda-3)(\lambda+5)(6-\lambda) = 0$. For $\lambda_1 = 3$, $\lambda_2 = -5$, and $\lambda_3 = 6$ we obtain

$$K_1 = \begin{pmatrix} 1 \\ 1 \\ 0 \end{pmatrix}, \quad K_2 = \begin{pmatrix} 1 \\ -1 \\ 0 \end{pmatrix}, \quad \text{and} \quad K_3 = \begin{pmatrix} 2 \\ -2 \\ 11 \end{pmatrix},$$

so that

$$X = c_1 \begin{pmatrix} 1 \\ 1 \\ 0 \end{pmatrix} e^{3t} + c_2 \begin{pmatrix} 1 \\ -1 \\ 0 \end{pmatrix} e^{-5t} + c_3 \begin{pmatrix} 2 \\ -2 \\ 11 \end{pmatrix} e^{6t}.$$

21. We have $\det(A - \lambda I) = (\lambda-2)^2 = 0$. For $\lambda_1 = 2$ we obtain

$$K = \begin{pmatrix} 1 \\ 1 \end{pmatrix}.$$

A solution of $(A - \lambda_1 I)P = K$ is

$$P = \begin{pmatrix} -1/3 \\ 0 \end{pmatrix}$$

so that

$$X = c_1 \begin{pmatrix} 1 \\ 1 \end{pmatrix} e^{2t} + c_2 \left[\begin{pmatrix} 1 \\ 1 \end{pmatrix} te^{2t} + \begin{pmatrix} -1/3 \\ 0 \end{pmatrix} e^{2t} \right].$$

24. We have $\det(A - \lambda I) = (\lambda-8)(\lambda+1)^2 = 0$. For $\lambda_1 = 8$ we obtain

$$K_1 = \begin{pmatrix} 2 \\ 1 \\ 2 \end{pmatrix}.$$

For $\lambda_2 = -1$ we obtain

$$\mathbf{K}_2 = \begin{pmatrix} 0 \\ -2 \\ 1 \end{pmatrix} \quad \text{and} \quad \mathbf{K}_3 = \begin{pmatrix} 1 \\ -2 \\ 0 \end{pmatrix}.$$

Then

$$\mathbf{X} = c_1 \begin{pmatrix} 2 \\ 1 \\ 2 \end{pmatrix} e^{8t} + c_2 \begin{pmatrix} 0 \\ -2 \\ 1 \end{pmatrix} e^{-t} + c_3 \begin{pmatrix} 1 \\ -2 \\ 0 \end{pmatrix} e^{-t}.$$

27. We have $\det(\mathbf{A} - \lambda\mathbf{I}) = -(\lambda - 1)^3 = 0$. For $\lambda_1 = 1$ we obtain

$$\mathbf{K} = \begin{pmatrix} 0 \\ 1 \\ 1 \end{pmatrix}.$$

Solutions of $(\mathbf{A} - \lambda_1\mathbf{I})\mathbf{P} = \mathbf{K}$ and $(\mathbf{A} - \lambda_1\mathbf{I})\mathbf{Q} = \mathbf{P}$ are

$$\mathbf{P} = \begin{pmatrix} 0 \\ 1 \\ 0 \end{pmatrix} \quad \text{and} \quad \mathbf{Q} = \begin{pmatrix} 1/2 \\ 0 \\ 0 \end{pmatrix}$$

so that

$$\mathbf{X} = c_1 \begin{pmatrix} 0 \\ 1 \\ 1 \end{pmatrix} + c_2 \left[\begin{pmatrix} 0 \\ 1 \\ 1 \end{pmatrix} te^t + \begin{pmatrix} 0 \\ 1 \\ 0 \end{pmatrix} e^t \right] + c_3 \left[\begin{pmatrix} 0 \\ 1 \\ 1 \end{pmatrix} \frac{t^2}{2} e^t + \begin{pmatrix} 0 \\ 1 \\ 0 \end{pmatrix} te^t + \begin{pmatrix} 1/2 \\ 0 \\ 0 \end{pmatrix} e^t \right].$$

30. We have $\det(\mathbf{A} - \lambda\mathbf{I}) = -(\lambda + 1)(\lambda - 1)^2 = 0$. For $\lambda_1 = -1$ we obtain

$$\mathbf{K}_1 = \begin{pmatrix} -1 \\ 0 \\ 1 \end{pmatrix}.$$

For $\lambda_2 = 1$ we obtain

$$\mathbf{K}_2 = \begin{pmatrix} 1 \\ 0 \\ 1 \end{pmatrix} \quad \text{and} \quad \mathbf{K}_3 = \begin{pmatrix} 0 \\ 1 \\ 0 \end{pmatrix}$$

so that

$$\mathbf{X} = c_1 \begin{pmatrix} -1 \\ 0 \\ 1 \end{pmatrix} e^{-t} + c_2 \begin{pmatrix} 1 \\ 0 \\ 1 \end{pmatrix} e^t + c_3 \begin{pmatrix} 0 \\ 1 \\ 0 \end{pmatrix} e^t.$$

If

$$\mathbf{X}(0) = \begin{pmatrix} 1 \\ 2 \\ 5 \end{pmatrix}$$

then $c_1 = 2$, $c_2 = 3$, and $c_3 = 2$.

In Problems 33-45 the form of the answer will vary according to the choice of eigenvector. For example, in Problem 33, if $\mathbf{K}_1$ is chosen to be $\begin{pmatrix} 1 \\ 2-i \end{pmatrix}$ the solution has the form

$$\mathbf{X} = c_1 \begin{pmatrix} \cos t \\ 2\cos t + \sin t \end{pmatrix} e^{4t} + c_2 \begin{pmatrix} \sin t \\ 2\sin t - \cos t \end{pmatrix} e^{4t}.$$

33. We have $\det(\mathbf{A} - \lambda\mathbf{I}) = \lambda^2 - 8\lambda + 17 = 0$. For $\lambda_1 = 4 + i$ we obtain

$$\mathbf{K}_1 = \begin{pmatrix} 2+i \\ 5 \end{pmatrix}$$

so that

$$\mathbf{X}_1 = \begin{pmatrix} 2+i \\ 5 \end{pmatrix} e^{(4+i)t} = \begin{pmatrix} 2\cos t - \sin t \\ 5\cos t \end{pmatrix} e^{4t} + i \begin{pmatrix} \cos t + 2\sin t \\ 5\sin t \end{pmatrix} e^{4t}.$$

Then

$$\mathbf{X} = c_1 \begin{pmatrix} 2\cos t - \sin t \\ 5\cos t \end{pmatrix} e^{4t} + c_2 \begin{pmatrix} 2\sin t + \cos t \\ 5\sin t \end{pmatrix} e^{4t}.$$

36. We have $\det(\mathbf{A} - \lambda\mathbf{I}) = \lambda^2 - 10\lambda + 34 = 0$. For $\lambda_1 = 5 + 3i$ we obtain

$$\mathbf{K}_1 = \begin{pmatrix} 1-3i \\ 2 \end{pmatrix}$$

so that

$$\mathbf{X}_1 = \begin{pmatrix} 1-3i \\ 2 \end{pmatrix} e^{(5+3i)t} = \begin{pmatrix} \cos 3t + 3\sin 3t \\ 2\cos 3t \end{pmatrix} e^{5t} + i \begin{pmatrix} \sin 3t - 3\cos 3t \\ 2\cos 3t \end{pmatrix} e^{5t}.$$

Then

$$\mathbf{X} = c_1 \begin{pmatrix} \cos 3t + 3\sin 3t \\ 2\cos 3t \end{pmatrix} e^{5t} + c_2 \begin{pmatrix} \sin 3t - 3\cos 3t \\ 2\cos 3t \end{pmatrix} e^{5t}.$$

39. We have $\det(\mathbf{A} - \lambda\mathbf{I}) = -\lambda\left(\lambda^2 + 1\right) = 0$. For $\lambda_1 = 0$ we obtain

$$\mathbf{K}_1 = \begin{pmatrix} 1 \\ 0 \\ 0 \end{pmatrix}.$$

For $\lambda_2 = i$ we obtain

$$\mathbf{K}_2 = \begin{pmatrix} -i \\ i \\ 1 \end{pmatrix}.$$

so that

$$\mathbf{X}_2 = \begin{pmatrix} -i \\ i \\ 1 \end{pmatrix} e^{it} = \begin{pmatrix} \sin t \\ -\sin t \\ \cos t \end{pmatrix} + i \begin{pmatrix} -\cos t \\ \cos t \\ \sin t \end{pmatrix}.$$

Then

$$\mathbf{X} = c_1 \begin{pmatrix} 1 \\ 0 \\ 0 \end{pmatrix} + c_2 \begin{pmatrix} \sin t \\ -\sin t \\ \cos t \end{pmatrix} + c_3 \begin{pmatrix} -\cos t \\ \cos t \\ \sin t \end{pmatrix}.$$

42. We have $\det(\mathbf{A} - \lambda\mathbf{I}) = -(\lambda - 6)(\lambda^2 - 8\lambda + 20) = 0$. For $\lambda_1 = 6$ we obtain

$$\mathbf{K}_1 = \begin{pmatrix} 0 \\ 1 \\ 0 \end{pmatrix}.$$

For $\lambda_2 = 4 + 2i$ we obtain

$$\mathbf{K}_2 = \begin{pmatrix} -i \\ 0 \\ 2 \end{pmatrix}$$

so that

$$\mathbf{X}_2 = \begin{pmatrix} -i \\ 0 \\ 2 \end{pmatrix} e^{(4+2i)t} = \begin{pmatrix} \sin 2t \\ 0 \\ 2\cos 2t \end{pmatrix} e^{4t} + i \begin{pmatrix} -\cos 2t \\ 0 \\ 2\sin 2t \end{pmatrix} e^{4t}.$$

Then

$$\mathbf{X} = c_1 \begin{pmatrix} 0 \\ 1 \\ 0 \end{pmatrix} e^{6t} + c_2 \begin{pmatrix} \sin 2t \\ 0 \\ 2\cos 2t \end{pmatrix} e^{4t} + c_3 \begin{pmatrix} -\cos 2t \\ 0 \\ 2\sin 2t \end{pmatrix} e^{4t}.$$

45. We have $\det(\mathbf{A} - \lambda\mathbf{I}) = (1 - \lambda)(\lambda^2 + 25) = 0$. For $\lambda_1 = 1$ we obtain

$$\mathbf{K}_1 = \begin{pmatrix} 25 \\ -7 \\ 6 \end{pmatrix}.$$

For $\lambda_2 = 5i$ we obtain

$$\mathbf{K}_2 = \begin{pmatrix} 1 + 5i \\ 1 \\ 1 \end{pmatrix}$$

so that

$$\mathbf{X}_2 = \begin{pmatrix} 1 + 5i \\ 1 \\ 1 \end{pmatrix} e^{5it} = \begin{pmatrix} \cos 5t - 5\sin 5t \\ \cos 5t \\ \cos 5t \end{pmatrix} + i \begin{pmatrix} \sin 5t + 5\cos 5t \\ \sin 5t \\ \sin 5t \end{pmatrix}.$$

Then

$$\mathbf{X} = c_1 \begin{pmatrix} 25 \\ -7 \\ 6 \end{pmatrix} e^t + c_2 \begin{pmatrix} \cos 5t - 5\sin 5t \\ \cos 5t \\ \cos 5t \end{pmatrix} + c_3 \begin{pmatrix} \sin 5t + 5\cos 5t \\ \sin 5t \\ \sin 5t \end{pmatrix}.$$

If

$$\mathbf{X}(0) = \begin{pmatrix} 4 \\ 6 \\ -7 \end{pmatrix}$$

then $c_1 = c_2 = -1$ and $c_3 = 6$.

Exercises 8.3

3. From

$$\mathbf{X}' = \begin{pmatrix} 3 & -5 \\ 3/4 & -1 \end{pmatrix} \mathbf{X} + \begin{pmatrix} 1 \\ -1 \end{pmatrix} e^{t/2}$$

we obtain

$$\mathbf{X}_c = c_1 \begin{pmatrix} 10 \\ 3 \end{pmatrix} e^{3t/2} + c_2 \begin{pmatrix} 2 \\ 1 \end{pmatrix} e^{t/2}.$$

Then

$$\mathbf{\Phi} = \begin{pmatrix} 10e^{3t/2} & 2e^{t/2} \\ 3e^{3t/2} & e^{t/2} \end{pmatrix} \quad \text{and} \quad \mathbf{\Phi}^{-1} = \begin{pmatrix} \frac{1}{4}e^{-3t/2} & -\frac{1}{2}e^{-3t/2} \\ -\frac{3}{4}e^{-t/2} & \frac{5}{2}e^{-t/2} \end{pmatrix}$$

so that

$$\mathbf{U} = \int \mathbf{\Phi}^{-1}\mathbf{F}\,dt = \int \begin{pmatrix} \frac{3}{4}e^{-t} \\ -\frac{13}{4} \end{pmatrix} dt = \begin{pmatrix} -\frac{3}{4}e^{-t} \\ -\frac{13}{4}t \end{pmatrix}$$

and

$$\mathbf{X}_p = \mathbf{\Phi}\mathbf{U} = \begin{pmatrix} -13/2 \\ -13/4 \end{pmatrix} te^{t/2} + \begin{pmatrix} -15/2 \\ -9/4 \end{pmatrix} e^{t/2}.$$

6. From

$$\mathbf{X}' = \begin{pmatrix} 0 & 2 \\ -1 & 3 \end{pmatrix} \mathbf{X} + \begin{pmatrix} 2 \\ e^{-3t} \end{pmatrix}$$

we obtain

$$\mathbf{X}_c = c_1 \begin{pmatrix} 2 \\ 1 \end{pmatrix} e^t + c_2 \begin{pmatrix} 1 \\ 1 \end{pmatrix} e^{2t}.$$

Then

$$\mathbf{\Phi} = \begin{pmatrix} 2e^t & e^{2t} \\ e^t & e^{2t} \end{pmatrix} \quad \text{and} \quad \mathbf{\Phi}^{-1} = \begin{pmatrix} e^{-t} & -e^{-t} \\ -e^{-2t} & 2e^{-2t} \end{pmatrix}$$

so that

$$\mathbf{U} = \int \mathbf{\Phi}^{-1}\mathbf{F}\,dt = \int \begin{pmatrix} 2e^{-t} - e^{-4t} \\ -2e^{-2t} + 2e^{-5t} \end{pmatrix} dt = \begin{pmatrix} -2e^{-t} + \frac{1}{4}e^{-4t} \\ e^{-2t} - \frac{2}{5}e^{-5t} \end{pmatrix}$$

and

$$\mathbf{X}_p = \mathbf{\Phi U} = \begin{pmatrix} \frac{1}{10}e^{-3t} - 3 \\ -\frac{3}{20}e^{-3t} - 1 \end{pmatrix}.$$

9. From

$$\mathbf{X}' = \begin{pmatrix} 3 & 2 \\ -2 & -1 \end{pmatrix} \mathbf{X} + \begin{pmatrix} 2 \\ 1 \end{pmatrix} e^{-t}$$

we obtain

$$\mathbf{X}_c = c_1 \begin{pmatrix} 1 \\ -1 \end{pmatrix} e^t + c_2 \left[\begin{pmatrix} 1 \\ -1 \end{pmatrix} te^t + \begin{pmatrix} 0 \\ 1/2 \end{pmatrix} e^t \right].$$

Then

$$\mathbf{\Phi} = \begin{pmatrix} e^t & te^t \\ -e^t & \frac{1}{2}e^t - te^t \end{pmatrix} \quad \text{and} \quad \mathbf{\Phi}^{-1} = \begin{pmatrix} e^{-t} - 2te^{-t} & -2te^{-t} \\ 2e^{-t} & 2e^{-t} \end{pmatrix}$$

so that

$$\mathbf{U} = \int \mathbf{\Phi}^{-1}\mathbf{F}\,dt = \int \begin{pmatrix} 2e^{-2t} - 6te^{-2t} \\ 6e^{-2t} \end{pmatrix} dt = \begin{pmatrix} \frac{1}{2}e^{-2t} + 3te^{-2t} \\ -3e^{-2t} \end{pmatrix}$$

and

$$\mathbf{X}_p = \mathbf{\Phi U} = \begin{pmatrix} 1/2 \\ -2 \end{pmatrix} e^{-t}.$$

12. From

$$\mathbf{X}' = \begin{pmatrix} 1 & -1 \\ 1 & 1 \end{pmatrix} \mathbf{X} + \begin{pmatrix} 3 \\ 3 \end{pmatrix} e^t$$

we obtain

$$\mathbf{X}_c = c_1 \begin{pmatrix} -\sin t \\ \cos t \end{pmatrix} e^t + c_2 \begin{pmatrix} \cos t \\ \sin t \end{pmatrix} e^t.$$

Then

$$\mathbf{\Phi} = \begin{pmatrix} -\sin t & \cos t \\ \cos t & \sin t \end{pmatrix} e^t \quad \text{and} \quad \mathbf{\Phi}^{-1} = \begin{pmatrix} -\sin t & \cos t \\ \cos t & \sin t \end{pmatrix} e^{-t}$$

so that

$$\mathbf{U} = \int \mathbf{\Phi}^{-1}\mathbf{F}\,dt = \int \begin{pmatrix} -3\sin t + 3\cos t \\ 3\cos t + 3\sin t \end{pmatrix} dt = \begin{pmatrix} 3\cos t + 3\sin t \\ 3\sin t - 3\cos t \end{pmatrix}$$

and

$$\mathbf{X}_p = \mathbf{\Phi U} = \begin{pmatrix} -3 \\ 3 \end{pmatrix} e^t.$$

15. From

$$\mathbf{X}' = \begin{pmatrix} 0 & 1 \\ -1 & 0 \end{pmatrix} \mathbf{X} + \begin{pmatrix} 0 \\ \sec t \tan t \end{pmatrix}$$

we obtain

$$\mathbf{X}_c = c_1 \begin{pmatrix} \cos t \\ -\sin t \end{pmatrix} + c_2 \begin{pmatrix} \sin t \\ \cos t \end{pmatrix}.$$

Then

$$\Phi = \begin{pmatrix} \cos t & \sin t \\ -\sin t & \cos t \end{pmatrix} t \quad \text{and} \quad \Phi^{-1} = \begin{pmatrix} \cos t & -\sin t \\ \sin t & \cos t \end{pmatrix}$$

so that

$$U = \int \Phi^{-1} F \, dt = \int \begin{pmatrix} -\tan^2 t \\ \tan t \end{pmatrix} dt = \begin{pmatrix} t - \tan t \\ \ln |\sec t| \end{pmatrix}$$

and

$$X_p = \Phi U = \begin{pmatrix} \cos t \\ -\sin t \end{pmatrix} t + \begin{pmatrix} -\sin t \\ \sin t \tan t \end{pmatrix} + \begin{pmatrix} \sin t \\ \cos t \end{pmatrix} \ln |\sec t|.$$

18. From

$$X' = \begin{pmatrix} 1 & -2 \\ 1 & -1 \end{pmatrix} X + \begin{pmatrix} \tan t \\ 1 \end{pmatrix}$$

we obtain

$$X_c = c_1 \begin{pmatrix} \cos t - \sin t \\ \cos t \end{pmatrix} + c_2 \begin{pmatrix} \cos t + \sin t \\ \sin t \end{pmatrix}.$$

Then

$$\Phi = \begin{pmatrix} \cos t - \sin t & \cos t + \sin t \\ \cos t & \sin t \end{pmatrix} \quad \text{and} \quad \Phi^{-1} = \begin{pmatrix} -\sin t & \cos t + \sin t \\ \cos t & \sin t - \cos t \end{pmatrix}$$

so that

$$U = \int \Phi^{-1} F \, dt = \int \begin{pmatrix} 2\cos t + \sin t - \sec t \\ 2\sin t - \cos t \end{pmatrix} dt = \begin{pmatrix} 2\sin t - \cos t - \ln |\sec t + \tan t| \\ -2\cos t - \sin t \end{pmatrix}$$

and

$$X_p = \Phi U = \begin{pmatrix} 3\sin t \cos t - \cos^2 t - 2\sin^2 t + (\sin t - \cos t) \ln |\sec t + \tan t| \\ \sin^2 t - \cos^2 t - \cos t (\ln |\sec t + \tan t|) \end{pmatrix}.$$

21. From

$$X' = \begin{pmatrix} 3 & -1 \\ -1 & 3 \end{pmatrix} X + \begin{pmatrix} 4e^{2t} \\ 4e^{4t} \end{pmatrix}$$

we obtain

$$\Phi = \begin{pmatrix} -e^{4t} & e^{2t} \\ e^{4t} & e^{2t} \end{pmatrix}, \quad \Phi^{-1} = \begin{pmatrix} -\frac{1}{2}e^{-4t} & \frac{1}{2}e^{-4t} \\ \frac{1}{2}e^{-2t} & \frac{1}{2}e^{-2t} \end{pmatrix},$$

and

$$X = \Phi \Phi^{-1}(0) X(0) + \Phi \int_0^t \Phi^{-1} F \, ds = \Phi \cdot \begin{pmatrix} 0 \\ 1 \end{pmatrix} + \Phi \cdot \begin{pmatrix} e^{-2t} + 2t - 1 \\ e^{2t} + 2t - 1 \end{pmatrix}$$

$$= \begin{pmatrix} 2 \\ 2 \end{pmatrix} te^{2t} + \begin{pmatrix} -1 \\ 1 \end{pmatrix} e^{2t} + \begin{pmatrix} -2 \\ 2 \end{pmatrix} te^{4t} + \begin{pmatrix} 2 \\ 0 \end{pmatrix} e^{4t}.$$

24. Solving

$$\begin{vmatrix} 6 - \lambda & 1 \\ 4 & 3 - \lambda \end{vmatrix} = \lambda^2 - 9\lambda + 14 = (\lambda - 2)(\lambda - 7) = 0$$

we obtain the eigenvalues $\lambda_1 = 2$ and $\lambda_2 = 7$. Corresponding eigenvectors are

$$\mathbf{K}_1 = \begin{pmatrix} 1 \\ -4 \end{pmatrix} \quad \text{and} \quad \mathbf{K}_2 = \begin{pmatrix} 1 \\ 1 \end{pmatrix}.$$

Thus

$$\mathbf{X}_c = c_1 \begin{pmatrix} 1 \\ -4 \end{pmatrix} e^{2t} + c_2 \begin{pmatrix} 1 \\ 1 \end{pmatrix} e^{7t}.$$

Substituting

$$\mathbf{X}_p = \begin{pmatrix} a_2 \\ b_2 \end{pmatrix} t + \begin{pmatrix} a_1 \\ b_1 \end{pmatrix}$$

into the system yields

$$6a_2 + b_2 + 6 = 0$$

$$4a_2 + 3b_2 - 10 = 0$$

$$6a_1 + b_1 - a_2 = 0$$

$$4a_1 + 3b_1 - b_2 + 4 = 0.$$

Solving the first two equations simultaneously yields $a_2 = -2$ and $b_2 = 6$. Substituting these two values into the second pair of equations and solving for a_1 and b_1 give $a_1 = -\frac{4}{7}$ and $b_1 = \frac{10}{7}$. Then

$$\mathbf{X}(t) = c_1 \begin{pmatrix} 1 \\ -4 \end{pmatrix} e^{2t} + c_2 \begin{pmatrix} 1 \\ 1 \end{pmatrix} e^{7t} + \begin{pmatrix} -2 \\ 6 \end{pmatrix} t + \begin{pmatrix} -\frac{4}{7} \\ \frac{10}{7} \end{pmatrix}.$$

27. Let $\mathbf{I} = \begin{pmatrix} i_1 \\ i_2 \end{pmatrix}$ so that

$$\mathbf{I}' = \begin{pmatrix} -11 & 3 \\ 3 & -3 \end{pmatrix} \mathbf{I} + \begin{pmatrix} 100 \sin t \\ 0 \end{pmatrix}$$

and

$$\mathbf{X}_c = c_1 \begin{pmatrix} 1 \\ 3 \end{pmatrix} e^{-2t} + c_2 \begin{pmatrix} 3 \\ -1 \end{pmatrix} e^{-12t}.$$

Then

$$\mathbf{\Phi} = \begin{pmatrix} e^{-2t} & 3e^{-12t} \\ 3e^{-2t} & -e^{-12t} \end{pmatrix}, \quad \mathbf{\Phi}^{-1} = \begin{pmatrix} \frac{1}{10}e^{2t} & \frac{3}{10}e^{2t} \\ \frac{3}{10}e^{12t} & -\frac{1}{10}e^{12t} \end{pmatrix},$$

$$\mathbf{U} = \int \mathbf{\Phi}^{-1}\mathbf{F}\, dt = \int \begin{pmatrix} 10e^{2t} \sin t \\ 30e^{12t} \sin t \end{pmatrix} dt = \begin{pmatrix} 2e^{2t}(2\sin t - \cos t) \\ \frac{6}{29}e^{12t}(12\sin t - \cos t) \end{pmatrix},$$

and

$$\mathbf{I}_p = \mathbf{\Phi}\mathbf{U} = \begin{pmatrix} \frac{332}{29}\sin t - \frac{76}{29}\cos t \\ \frac{276}{29}\sin t - \frac{168}{29}\cos t \end{pmatrix}$$

so that

$$\mathbf{I} = c_1 \begin{pmatrix} 1 \\ 3 \end{pmatrix} e^{-2t} + c_2 \begin{pmatrix} 3 \\ -1 \end{pmatrix} e^{-12t} + \mathbf{I}_p.$$

If $\mathbf{I}(0) = \begin{pmatrix} 0 \\ 0 \end{pmatrix}$ then $c_1 = 2$ and $c_2 = \frac{6}{29}$.

Exercises 8.4

3. For

$$\mathbf{A} = \begin{pmatrix} 1 & 1 & 1 \\ 1 & 1 & 1 \\ -2 & -2 & -2 \end{pmatrix}$$

we have

$$\mathbf{A}^2 = \begin{pmatrix} 1 & 1 & 1 \\ 1 & 1 & 1 \\ -2 & -2 & -2 \end{pmatrix} \begin{pmatrix} 1 & 1 & 1 \\ 1 & 1 & 1 \\ -2 & -2 & -2 \end{pmatrix} = \begin{pmatrix} 0 & 0 & 0 \\ 0 & 0 & 0 \\ 0 & 0 & 0 \end{pmatrix}.$$

Thus, $\mathbf{A}^3 = \mathbf{A}^4 = \mathbf{A}^5 = \cdots = \mathbf{0}$ and

$$e^{\mathbf{A}t} = \mathbf{I} + \mathbf{A}t = \begin{pmatrix} 1 & 0 & 0 \\ 0 & 1 & 0 \\ 0 & 0 & 1 \end{pmatrix} + \begin{pmatrix} t & t & t \\ t & t & t \\ -2t & -2t & -2t \end{pmatrix} = \begin{pmatrix} t+1 & t & t \\ t & t+1 & t \\ -2t & -2t & -2t+1 \end{pmatrix}.$$

6. In Problem 2 it is shown that

$$e^{\mathbf{A}t} = \begin{pmatrix} \cosh t & \sinh t \\ \sinh t & \cosh t \end{pmatrix}.$$

Thus

$$\mathbf{X} = \begin{pmatrix} \cosh t & \sinh t \\ \sinh t & \cosh t \end{pmatrix} \begin{pmatrix} c_1 \\ c_2 \end{pmatrix} = c_1 \begin{pmatrix} \cosh t \\ \sinh t \end{pmatrix} + c_2 \begin{pmatrix} \sinh t \\ \cosh t \end{pmatrix}.$$

9. In Problem 1 it is shown that

$$e^{\mathbf{A}t} = \begin{pmatrix} e^t & 0 \\ 0 & e^{2t} \end{pmatrix} \quad \text{and} \quad e^{-\mathbf{A}t} = \begin{pmatrix} e^{-t} & 0 \\ 0 & e^{-2t} \end{pmatrix}.$$

To solve

$$\mathbf{X}' = \begin{pmatrix} 1 & 0 \\ 0 & 2 \end{pmatrix} \mathbf{X} + \begin{pmatrix} 3 \\ -1 \end{pmatrix}$$

we identify $t_0 = 0$, $\mathbf{F}(s) = \begin{pmatrix} 3 \\ -1 \end{pmatrix}$, and use the results of Problem 1 and equation (5) in the text.

$$\mathbf{X}(t) = e^{\mathbf{A}t}\mathbf{C} + e^{\mathbf{A}t}\int_{t_0}^{t} e^{-\mathbf{A}s}\mathbf{F}(s)\,ds$$

$$= \begin{pmatrix} e^t & 0 \\ 0 & e^{2t} \end{pmatrix}\begin{pmatrix} c_1 \\ c_2 \end{pmatrix} + \begin{pmatrix} e^t & 0 \\ 0 & e^{2t} \end{pmatrix}\int_0^t \begin{pmatrix} e^{-s} & 0 \\ 0 & e^{-2s} \end{pmatrix}\begin{pmatrix} 3 \\ -1 \end{pmatrix} ds$$

$$= \begin{pmatrix} c_1 e^t \\ c_2 e^{2t} \end{pmatrix} + \begin{pmatrix} e^t & 0 \\ 0 & e^{2t} \end{pmatrix}\int_0^t \begin{pmatrix} 3e^{-s} \\ -e^{-2s} \end{pmatrix} ds$$

$$= \begin{pmatrix} c_1 e^t \\ c_2 e^{2t} \end{pmatrix} + \begin{pmatrix} e^t & 0 \\ 0 & e^{2t} \end{pmatrix}\begin{pmatrix} -3e^{-s} \\ \frac{1}{2}e^{-2s} \end{pmatrix}\Big|_0^t$$

$$= \begin{pmatrix} c_1 e^t \\ c_2 e^{2t} \end{pmatrix} + \begin{pmatrix} e^t & 0 \\ 0 & e^{2t} \end{pmatrix}\begin{pmatrix} -3e^{-t} - 3 \\ \frac{1}{2}e^{-2t} - \frac{1}{2} \end{pmatrix}$$

$$= \begin{pmatrix} c_1 e^t \\ c_2 e^{2t} \end{pmatrix} + \begin{pmatrix} -3 - 3e^t \\ \frac{1}{2} - \frac{1}{2}e^{2t} \end{pmatrix} = c_3 \begin{pmatrix} 1 \\ 0 \end{pmatrix} e^t + c_4 \begin{pmatrix} 0 \\ 1 \end{pmatrix} e^{2t} + \begin{pmatrix} -3 \\ \frac{1}{2} \end{pmatrix}.$$

12. In Problem 2 it is shown that

$$e^{\mathbf{A}t} = \begin{pmatrix} \cosh t & \sinh t \\ \sinh t & \cosh t \end{pmatrix} \quad \text{and} \quad e^{-\mathbf{A}t} = \begin{pmatrix} \cosh t & -\sinh t \\ -\sinh t & \cosh t \end{pmatrix}.$$

To solve

$$\mathbf{X}' = \begin{pmatrix} 0 & 1 \\ 1 & 0 \end{pmatrix}\mathbf{X} + \begin{pmatrix} \cosh t \\ \sinh t \end{pmatrix}$$

we identify $t_0 = 0$, $\mathbf{F}(s) = \begin{pmatrix} \cosh t \\ \sinh t \end{pmatrix}$, and use the results of Problem 2 and equation (5) in the text.

$$\mathbf{X}(t) = e^{\mathbf{A}t}\mathbf{C} + e^{\mathbf{A}t}\int_{t_0}^{t} e^{-\mathbf{A}s}\mathbf{F}(s)\,ds$$

$$= \begin{pmatrix} \cosh t & \sinh t \\ \sinh t & \cosh t \end{pmatrix}\begin{pmatrix} c_1 \\ c_2 \end{pmatrix} + \begin{pmatrix} \cosh t & \sinh t \\ \sinh t & \cosh t \end{pmatrix}\int_0^t \begin{pmatrix} \cosh s & -\sinh s \\ -\sinh s & \cosh s \end{pmatrix}\begin{pmatrix} \cosh s \\ \sinh s \end{pmatrix} ds$$

$$= \begin{pmatrix} c_1 \cosh t + c_2 \sinh t \\ c_1 \sinh t + c_2 \cosh t \end{pmatrix} + \begin{pmatrix} \cosh t & \sinh t \\ \sinh t & \cosh t \end{pmatrix}\int_0^t \begin{pmatrix} 1 \\ 0 \end{pmatrix} ds$$

$$= \begin{pmatrix} c_1 \cosh t + c_2 \sinh t \\ c_1 \sinh t + c_2 \cosh t \end{pmatrix} + \begin{pmatrix} \cosh t & \sinh t \\ \sinh t & \cosh t \end{pmatrix}\begin{pmatrix} s \\ 0 \end{pmatrix}\Big|_0^t$$

$$= \begin{pmatrix} c_1 \cosh t + c_2 \sinh t \\ c_1 \sinh t + c_2 \cosh t \end{pmatrix} + \begin{pmatrix} \cosh t & \sinh t \\ \sinh t & \cosh t \end{pmatrix} \begin{pmatrix} t \\ 0 \end{pmatrix}$$

$$= \begin{pmatrix} c_1 \cosh t + c_2 \sinh t \\ c_1 \sinh t + c_2 \cosh t \end{pmatrix} + \begin{pmatrix} t \cosh t \\ t \sinh t \end{pmatrix} = c_1 \begin{pmatrix} \cosh t \\ \sinh t \end{pmatrix} + c_2 \begin{pmatrix} \sinh t \\ \cosh t \end{pmatrix} + t \begin{pmatrix} \cosh t \\ \sinh t \end{pmatrix}.$$

15. From $s\mathbf{I} - \mathbf{A} = \begin{pmatrix} s-4 & -3 \\ 4 & s+4 \end{pmatrix}$ we find

$$(s\mathbf{I} - \mathbf{A})^{-1} = \begin{pmatrix} \dfrac{3/2}{s-2} - \dfrac{1/2}{s+2} & \dfrac{3/4}{s-2} - \dfrac{3/4}{s+2} \\[3mm] \dfrac{-1}{s-2} + \dfrac{1}{s+2} & \dfrac{-1/2}{s-2} + \dfrac{3/2}{s+2} \end{pmatrix}$$

and

$$e^{\mathbf{A}t} = \begin{pmatrix} \frac{3}{2}e^{2t} - \frac{1}{2}e^{-2t} & \frac{3}{4}e^{2t} - \frac{3}{4}e^{-2t} \\[2mm] -e^{2t} + e^{-2t} & -\frac{1}{2}e^{2t} + \frac{3}{2}e^{-2t} \end{pmatrix}.$$

The general solution of the system is then

$$\mathbf{X} = e^{\mathbf{A}t}\mathbf{C} = \begin{pmatrix} \frac{3}{2}e^{2t} - \frac{1}{2}e^{-2t} & \frac{3}{4}e^{2t} - \frac{3}{4}e^{-2t} \\[2mm] -e^{2t} + e^{-2t} & -\frac{1}{2}e^{2t} + \frac{3}{2}e^{-2t} \end{pmatrix} \begin{pmatrix} c_1 \\ c_2 \end{pmatrix}$$

$$= c_1 \begin{pmatrix} 3/2 \\ -1 \end{pmatrix} e^{2t} + c_1 \begin{pmatrix} -1/2 \\ 1 \end{pmatrix} e^{-2t} + c_2 \begin{pmatrix} 3/4 \\ -1/2 \end{pmatrix} e^{2t} + c_2 \begin{pmatrix} -3/4 \\ 3/2 \end{pmatrix} e^{-2t}$$

$$= \left(\frac{1}{2}c_1 + \frac{1}{4}c_2\right) \begin{pmatrix} 3 \\ -2 \end{pmatrix} e^{2t} + \left(-\frac{1}{2}c_1 - \frac{3}{4}c_2\right) \begin{pmatrix} 1 \\ -2 \end{pmatrix} e^{-2t}$$

$$= c_3 \begin{pmatrix} 3 \\ -2 \end{pmatrix} e^{2t} + c_4 \begin{pmatrix} 1 \\ -2 \end{pmatrix} e^{-2t}.$$

18. From $s\mathbf{I} - \mathbf{A} = \begin{pmatrix} s & -1 \\ 2 & s+2 \end{pmatrix}$ we find

$$(s\mathbf{I} - \mathbf{A})^{-1} = \begin{pmatrix} \dfrac{s+1+1}{(s+1)^2+1} & \dfrac{1}{(s+1)^2+1} \\[4mm] \dfrac{-2}{(s+1)^2+1} & \dfrac{s+1+1}{(s+1)^2+1} \end{pmatrix}$$

and

$$e^{\mathbf{A}t} = \begin{pmatrix} e^{-t}\cos t + e^{-t}\sin t & e^{-t}\sin t \\[2mm] -2e^{-t}\sin t & e^{-t}\cos t - e^{-t}\sin t \end{pmatrix}.$$

The general solution of the system is then

$$\mathbf{X} = e^{\mathbf{A}t}\mathbf{C} = \begin{pmatrix} e^{-t}\cos t + e^{-t}\sin t & e^{-t}\sin t \\ -2e^{-t}\sin t & e^{-t}\cos t - e^{-t}\sin t \end{pmatrix} \begin{pmatrix} c_1 \\ c_2 \end{pmatrix}$$

$$= c_1 \begin{pmatrix} 1 \\ 0 \end{pmatrix} e^{-t}\cos t + c_1 \begin{pmatrix} 1 \\ -2 \end{pmatrix} e^{-t}\sin t + c_2 \begin{pmatrix} 0 \\ 1 \end{pmatrix} e^{-t}\cos t + c_2 \begin{pmatrix} 1 \\ -1 \end{pmatrix} e^{-t}\sin t$$

$$= c_1 \begin{pmatrix} \cos t + \sin t \\ -2\sin t \end{pmatrix} e^{-t} + c_2 \begin{pmatrix} \sin t \\ \cos t - \sin t \end{pmatrix} e^{-t}.$$

21. From equation (3) in the text

$$e^{t\mathbf{A}} = e^{t\mathbf{PDP}^{-1}} = \mathbf{I} + t(\mathbf{PDP}^{-1}) + \frac{1}{2!}t^2(\mathbf{PDP}^{-1})^2 + \frac{1}{3!}t^3(\mathbf{PDP}^{-1})^3 + \cdots$$

$$= \mathbf{P}\left[\mathbf{I} + t\mathbf{D} + \frac{1}{2!}(t\mathbf{D})^2 + \frac{1}{3!}(t\mathbf{D})^3 + \cdots\right]\mathbf{P}^{-1} = \mathbf{P}e^{t\mathbf{D}}\mathbf{P}^{-1}.$$

24. From Problems 20-22 and equation (1) in the text

$$\mathbf{X} = e^{t\mathbf{A}}\mathbf{C} = \mathbf{P}e^{t\mathbf{D}}\mathbf{P}^{-1}\mathbf{C}$$

$$= \begin{pmatrix} -e^t & e^{3t} \\ e^t & e^{3t} \end{pmatrix} \begin{pmatrix} e^t & 0 \\ 0 & e^{3t} \end{pmatrix} \begin{pmatrix} -\frac{1}{2}e^{-t} & \frac{1}{2}e^{-t} \\ \frac{1}{2}e^{3t} & \frac{1}{2}e^{-3t} \end{pmatrix} \begin{pmatrix} c_1 \\ c_2 \end{pmatrix}$$

$$= \begin{pmatrix} \frac{1}{2}e^t + \frac{1}{2}e^{9t} & -\frac{1}{2}e^t + \frac{1}{2}e^{3t} \\ -\frac{1}{2}e^t + \frac{1}{2}e^{9t} & \frac{1}{2}e^t + \frac{1}{2}e^{3t} \end{pmatrix} \begin{pmatrix} c_1 \\ c_2 \end{pmatrix}.$$

———— Chapter 8 Review Exercises ————

3. Since

$$\begin{pmatrix} 4 & 6 & 6 \\ 1 & 3 & 2 \\ -1 & -4 & -3 \end{pmatrix} \begin{pmatrix} 3 \\ 1 \\ -1 \end{pmatrix} = \begin{pmatrix} 12 \\ 4 \\ -4 \end{pmatrix} = 4 \begin{pmatrix} 3 \\ 1 \\ -1 \end{pmatrix},$$

we see that $\lambda = 4$ is an eigenvalue with eigenvector $\mathbf{K}_3$. The corresponding solution is $\mathbf{X} = \mathbf{K}_3 e^{4t}$.

6. We have $\det(\mathbf{A} - \lambda\mathbf{I}) = (\lambda + 6)(\lambda + 2) = 0$ so that

$$\mathbf{X} = c_1 \begin{pmatrix} 1 \\ -1 \end{pmatrix} e^{-6t} + c_2 \begin{pmatrix} 1 \\ 1 \end{pmatrix} e^{-2t}.$$

9. We have $\det(\mathbf{A} - \lambda\mathbf{I}) = -(\lambda - 2)(\lambda - 4)(\lambda + 3) = 0$ so that

$$\mathbf{X} = c_1 \begin{pmatrix} -2 \\ 3 \\ 1 \end{pmatrix} e^{2t} + c_2 \begin{pmatrix} 0 \\ 1 \\ 1 \end{pmatrix} e^{4t} + c_3 \begin{pmatrix} 7 \\ 12 \\ -16 \end{pmatrix} e^{-3t}.$$

12. We have

$$\mathbf{X}_c = c_1 \begin{pmatrix} 2\cos t \\ -\sin t \end{pmatrix} e^t + c_2 \begin{pmatrix} 2\sin t \\ \cos t \end{pmatrix} e^t.$$

Then

$$\boldsymbol{\Phi} = \begin{pmatrix} 2\cos t & 2\sin t \\ -\sin t & \cos t \end{pmatrix} e^t, \quad \boldsymbol{\Phi}^{-1} = \begin{pmatrix} \frac{1}{2}\cos t & -\sin t \\ \frac{1}{2}\sin t & \cos t \end{pmatrix} e^{-t},$$

and

$$\mathbf{U} = \int \boldsymbol{\Phi}^{-1}\mathbf{F}\,dt = \int \begin{pmatrix} \cos t - \sec t \\ \sin t \end{pmatrix} dt = \begin{pmatrix} \sin t - \ln|\sec t + \tan t| \\ -\cos t \end{pmatrix},$$

so that

$$\mathbf{X}_p = \boldsymbol{\Phi}\mathbf{U} = \begin{pmatrix} -2\cos t\ln|\sec t + \tan t| \\ -1 + \sin t\ln|\sec t + \tan t| \end{pmatrix}.$$

15. (a) Letting

$$\mathbf{K} = \begin{pmatrix} k_1 \\ k_2 \\ k_3 \end{pmatrix}$$

we note that $(\mathbf{A} - 2\mathbf{I})\mathbf{K} = \mathbf{0}$ implies that $3k_1 + 3k_2 + 3k_3 = 0$, so $k_1 = -(k_2 + k_3)$. Choosing $k_2 = 0$, $k_3 = 1$ and then $k_2 = 1$, $k_3 = 0$ we get

$$\mathbf{K}_1 = \begin{pmatrix} -1 \\ 0 \\ 1 \end{pmatrix} \quad \text{and} \quad \mathbf{K}_2 = \begin{pmatrix} -1 \\ 1 \\ 0 \end{pmatrix},$$

respectively. Thus,

$$\mathbf{X}_1 = \begin{pmatrix} -1 \\ 0 \\ 1 \end{pmatrix} e^{2t} \quad \text{and} \quad \mathbf{X}_2 = \begin{pmatrix} -1 \\ 1 \\ 0 \end{pmatrix} e^{2t}$$

are two solutions.

(b) From $\det(\mathbf{A} - \lambda\mathbf{I}) = \lambda^2(3 - \lambda) = 0$ we see that $\lambda_1 = 3$, and 0 is an eigenvalue of multiplicity two. Letting

$$\mathbf{K} = \begin{pmatrix} k_1 \\ k_2 \\ k_3 \end{pmatrix},$$

as in part **(a)**, we note that $(\mathbf{A} - 0\mathbf{I})\mathbf{K} = \mathbf{A}\mathbf{K} = \mathbf{0}$ implies that $k_1 + k_2 + k_3 = 0$, so $k_1 = -(k_2 + k_3)$. Choosing $k_2 = 0$, $k_3 = 1$, and then $k_2 = 1$, $k_3 = 0$ we get

$$\mathbf{K}_2 = \begin{pmatrix} -1 \\ 0 \\ 1 \end{pmatrix} \quad \text{and} \quad \mathbf{K}_3 = \begin{pmatrix} -1 \\ 1 \\ 0 \end{pmatrix},$$

respectively. Since the eigenvector corresponding to $\lambda_1 = 3$ is

$$\mathbf{K}_1 \begin{pmatrix} 1 \\ 1 \\ 1 \end{pmatrix},$$

the general solution of the system is

$$\mathbf{X} = c_1 \begin{pmatrix} 1 \\ 1 \\ 1 \end{pmatrix} e^{3t} + c_2 \begin{pmatrix} -1 \\ 0 \\ 1 \end{pmatrix} + c_3 \begin{pmatrix} -1 \\ 1 \\ 0 \end{pmatrix}.$$

9 Numerical Solutions of Differential Equations

The tables in this chapter were constructed in a spreadsheet program which does not support subscripts. Consequently, x_n and y_n will be indicated as $x(n)$ and $y(n)$, respectively.

3.

$h = 0.1$		$h = 0.05$	
$x(n)$	$y(n)$	$x(n)$	$y(n)$
0.00	0.0000	0.00	0.0000
0.10	0.1005	0.05	0.0501
0.20	0.2030	0.10	0.1004
0.30	0.3098	0.15	0.1512
0.40	0.4234	0.20	0.2028
0.50	0.5470	0.25	0.2554
		0.30	0.3095
		0.35	0.3652
		0.40	0.4230
		0.45	0.4832
		0.50	0.5465

6.

$h = 0.1$		$h = 0.05$	
$x(n)$	$y(n)$	$x(n)$	$y(n)$
0.00	0.0000	0.00	0.0000
0.10	0.0050	0.05	0.0013
0.20	0.0200	0.10	0.0050
0.30	0.0451	0.15	0.0113
0.40	0.0805	0.20	0.0200
0.50	0.1266	0.25	0.0313
		0.30	0.0451
		0.35	0.0615
		0.40	0.0805
		0.45	0.1022
		0.50	0.1266

9.

$h = 0.1$		$h = 0.05$	
$x(n)$	$y(n)$	$x(n)$	$y(n)$
1.00	1.0000	1.00	1.0000
1.10	1.0095	1.05	1.0024
1.20	1.0404	1.10	1.0100
1.30	1.0967	1.15	1.0228
1.40	1.1866	1.20	1.0414
1.50	1.3260	1.25	1.0663
		1.30	1.0984
		1.35	1.1389
		1.40	1.1895
		1.45	1.2526
		1.50	1.3315

12. (a)

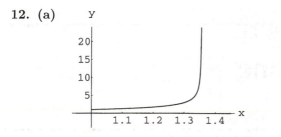

(b)

h=0.1	EULER	IMPROVED EULER
x(n)	y(n)	y(n)
1.00	1.0000	1.0000
1.10	1.2000	1.2469
1.20	1.4938	1.6668
1.30	1.9711	2.6427
1.40	2.9060	8.7988

15. (a) Using the Euler method we obtain $y(0.1) \approx y_1 = 0.8$.

(b) Using $y'' = 5e^{-2x}$ we see that the local truncation error is

$$5e^{-2c}\frac{(0.1)^2}{2} = 0.025e^{-2c}.$$

Since e^{-2x} is a decreasing function, $e^{-2c} \le e^0 = 1$ for $0 \le c \le 0.1$. Thus an upper bound for the local truncation error is $0.025(1) = 0.025$.

(c) Since $y(0.1) = 0.8234$, the actual error is $y(0.1) - y_1 = 0.0234$, which is less than 0.025.

(d) Using the Euler method with $h = 0.05$ we obtain $y(0.1) \approx y_2 = 0.8125$.

(e) The error in (d) is $0.8234 - 0.8125 = 0.0109$. With global truncation error $O(h)$, when the step size is halved we expect the error for $h = 0.05$ to be one-half the error when $h = 0.1$. Comparing 0.0109 with 0.0234 we see that this is the case.

18. (a) Using $y''' = -114e^{-3(x-1)}$ we see that the local truncation error is

$$\left| y'''(c)\frac{h^3}{6} \right| = 114e^{-3(x-1)}\frac{h^3}{6} = 19h^3 e^{-3(c-1)}.$$

(b) Since $e^{-3(x-1)}$ is a decreasing function for $1 \le x \le 1.5$, $e^{-3(c-1)} \le e^{-3(1-1)} = 1$ for $1 \le c \le 1.5$ and

$$\left| y'''(c)\frac{h^3}{6} \right| \le 19(0.1)^3(1) = 0.019.$$

(c) Using the improved Euler method with $h = 0.1$ we obtain $y(1.5) \approx 2.080108$. With $h = 0.05$ we obtain $y(1.5) \approx 2.059166$.

(d) Since $y(1.5) = 2.053216$, the error for $h = 0.1$ is $E_{0.1} = 0.026892$, while the error for $h = 0.05$ is $E_{0.05} = 0.005950$. With global truncation error $O(h^2)$ we expect $E_{0.1}/E_{0.05} \approx 4$. We actually have $E_{0.1}/E_{0.05} = 4.52$.

Exercises 9.2

3.

x(n)	y(n)
1.00	5.0000
1.10	3.9724
1.20	3.2284
1.30	2.6945
1.40	2.3163
1.50	2.0533

6.

x(n)	y(n)
0.00	1.0000
0.10	1.1115
0.20	1.2530
0.30	1.4397
0.40	1.6961
0.50	2.0670

9.

x(n)	y(n)
0.00	0.5000
0.10	0.5213
0.20	0.5358
0.30	0.5443
0.40	0.5482
0.50	0.5493

12.

x(n)	y(n)
0.00	0.5000
0.10	0.5250
0.20	0.5498
0.30	0.5744
0.40	0.5987
0.50	0.6225

15. (a)

x(n)	h = 0.05 y(n)	h = 0.1 y(n)
1.00	1.0000	1.0000
1.05	1.1112	
1.10	1.2511	1.2511
1.15	1.4348	
1.20	1.6934	1.6934
1.25	2.1047	
1.30	2.9560	2.9425
1.35	7.8981	
1.40	1.06E+15	903.0282

(b)

18. (a) Using $y^{(5)} = -1026e^{-3(x-1)}$ we see that the local truncation error is

$$\left| y^{(5)}(c) \frac{h^5}{120} \right| = 8.55h^5 e^{-3(c-1)}.$$

(b) Since $e^{-3(x-1)}$ is a decreasing function for $1 \le x \le 1.5$, $e^{-3(c-1)} \le e^{-3(1-1)} = 1$ for $1 \le c \le 1.5$ and

$$y^{(5)}(c) \frac{h^5}{120} \le 8.55(0.1)^5(1) = 0.0000855.$$

(c) Using the fourth-order Runge-Kutta method with $h = 0.1$ we obtain $y(1.5) \approx 2.053338827$. With $h = 0.05$ we obtain $y(1.5) \approx 2.053222989$.

Exercises 9.3

3.

x(n)	y(n)	
0.00	1.0000	initial condition
0.20	0.7328	Runge-Kutta
0.40	0.6461	Runge-Kutta
0.60	0.6585	Runge-Kutta
	0.7332	*predictor*
0.80	0.7232	corrector

6.

x(n)	y(n)	
0.00	1.0000	initial condition
0.20	1.4414	Runge-Kutta
0.40	1.9719	Runge-Kutta
0.60	2.6028	Runge-Kutta
	3.3483	*predictor*
0.80	3.3486	corrector
	4.2276	*predictor*
1.00	4.2280	corrector

x(n)	y(n)	
0.00	1.0000	initial condition
0.10	1.2102	Runge-Kutta
0.20	1.4414	Runge-Kutta
0.30	1.6949	Runge-Kutta
	1.9719	*predictor*
0.40	1.9719	corrector
	2.2740	*predictor*
0.50	2.2740	corrector
	2.6028	*predictor*
0.60	2.6028	corrector
	2.9603	*predictor*
0.70	2.9603	corrector
	3.3486	*predictor*
0.80	3.3486	corrector
	3.7703	*predictor*
0.90	3.7703	corrector
	4.2280	*predictor*
1.00	4.2280	corrector

Exercises 9.4

3. The substitution $y' = u$ leads to the system

$$y' = u, \qquad u' = 4u - 4y.$$

Using formula (4) in the text with x corresponding to t, y corresponding to x, and u corresponding to y, we obtain

								Runge-Kutta method with h=0.2		
m1	*m2*	*m3*	*m4*	*k1*	*k2*	*k3*	*k4*	*x*	*y*	*u*
								0.00	-2.0000	1.0000
0.2000	0.4400	0.5280	0.9072	2.4000	3.2800	3.5360	4.8064	0.20	-1.4928	4.4731

								Runge-Kutta method with h=0.1		
m1	*m2*	*m3*	*m4*	*k1*	*k2*	*k3*	*k4*	*x*	*y*	*u*
								0.00	-2.0000	1.0000
0.1000	0.1600	0.1710	0.2452	1.2000	1.4200	1.4520	1.7124	0.10	-1.8321	2.4427
0.2443	0.3298	0.3444	0.4487	1.7099	2.0031	2.0446	2.3900	0.20	-1.4919	4.4753

6.

								Runge-Kutta method with h=0.1		
m1	*m2*	*m3*	*m4*	*k1*	*k2*	*k3*	*k4*	*t*	*i1*	*i2*
								0.00	0.0000	0.0000
10.0000	0.0000	12.5000	-20.0000	0.0000	5.0000	-5.0000	22.5000	0.10	2.5000	3.7500
8.7500	-2.5000	13.4375	-28.7500	-5.0000	4.3750	-10.6250	29.6875	0.20	2.8125	5.7813
10.1563	-4.3750	17.0703	-40.0000	-8.7500	5.0781	-16.0156	40.3516	0.30	2.0703	7.4023
13.2617	-6.3672	22.9443	-55.1758	-12.7344	6.6309	-22.5488	55.3076	0.40	0.6104	9.1919
17.9712	-8.8867	31.3507	-75.9326	-17.7734	8.9856	-31.2024	75.9821	0.50	-1.5619	11.4877

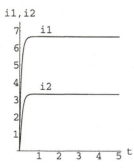

As $t \to \infty$ we see that $i_1(t) \to 6.75$ and $i_2(t) \to 3.4$.

9.

m1	m2	m3	m4	k1	k2	k3	k4	t	x	y
								0.00	-3.0000	5.0000
-1.0000	-0.9200	-0.9080	-0.8176	-0.6000	-0.7200	-0.7120	-0.8216	0.20	-3.9123	4.2857

m1	m2	m3	m4	k1	k2	k3	k4	t	x	y
								0.00	-3.0000	5.0000
-0.5000	-0.4800	-0.4785	-0.4571	-0.3000	-0.3300	-0.3290	-0.3579	0.10	-3.4790	4.6707
-0.4571	-0.4342	-0.4328	-0.4086	-0.3579	-0.3858	-0.3846	-0.4112	0.20	-3.9123	4.2857

12. Solving for x' and y' we obtain the system

$$x' = \frac{1}{2}y - 3t^2 + 2t - 5$$

$$y' = -\frac{1}{2}y + 3t^2 + 2t + 5.$$

m1	m2	m3	m4	k1	k2	k3	k4	t	x	y
								0.00	3.0000	-1.0000
-1.1000	-1.0110	-1.0115	-0.9349	1.1000	1.0910	1.0915	1.0949	0.20	1.9867	0.0933

m1	m2	m3	m4	k1	k2	k3	k4	t	x	y
								0.00	3.0000	-1.0000
-0.5500	-0.5270	-0.5271	-0.5056	0.5500	0.5470	0.5471	0.5456	0.10	2.4727	-0.4527
-0.5056	-0.4857	-0.4857	-0.4673	0.5456	0.5457	0.5457	0.5473	0.20	1.9867	0.0933

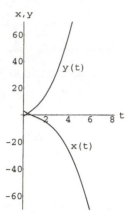

—————— **Exercises 9.5** ——————

3. We identify $P(x) = 2$, $Q(x) = 1$, $f(x) = 5x$, and $h = (1 - 0)/5 = 0.2$. Then the finite difference equation is

$$1.2y_{i+1} - 1.96y_i + 0.8y_{i-1} = 0.04(5x_i).$$

The solution of the corresponding linear system gives

x	0.0	0.2	0.4	0.6	0.8	1.0
y	0.0000	-0.2259	-0.3356	-0.3308	-0.2167	0.0000

6. We identify $P(x) = 5$, $Q(x) = 0$, $f(x) = 4\sqrt{x}$, and $h = (2 - 1)/6 = 0.1667$. Then the finite difference equation is

$$1.4167y_{i+1} - 2y_i + 0.5833y_{i-1} = 0.2778(4\sqrt{x_i}).$$

The solution of the corresponding linear system gives

x	1.0000	1.1667	1.3333	1.5000	1.6667	1.8333	2.0000
y	1.0000	-0.5918	-1.1626	-1.3070	-1.2704	-1.1541	-1.0000

9. We identify $P(x) = 1 - x$, $Q(x) = x$, $f(x) = x$, and $h = (1 - 0)/10 = 0.1$. Then the finite difference equation is

$$[1 + 0.05(1 - x_i)]y_{i+1} + [-2 + 0.01x_i]y_i + [1 - 0.05(1 - x_i)]y_{i-1} = 0.01x_i.$$

The solution of the corresponding linear system gives

x	0.0	0.1	0.2	0.3	0.4	0.5	0.6
y	0.0000	0.2660	0.5097	0.7357	0.9471	1.1465	1.3353

	0.7	0.8	0.9	1.0
	1.5149	1.6855	1.8474	2.0000

12. We identify $P(r) = 2/r$, $Q(r) = 0$, $f(r) = 0$, and $h = (4-1)/6 = 0.5$. Then the finite difference equation is

$$\left(1 + \frac{0.5}{r_i}\right) u_{i+1} - 2u_i + \left(1 - \frac{0.5}{r_i}\right) u_{i-1} = 0.$$

The solution of the corresponding linear system gives

r	1.0	1.5	2.0	2.5	3.0	3.5	4.0
u	50.0000	72.2222	83.3333	90.0000	94.4444	97.6190	100.0000

Chapter 9 Review Exercises

3.

h=0.1		IMPROVED	RUNGE	h=0.05		IMPROVED	RUNGE
$x(n)$	EULER	EULER	KUTTA	$x(n)$	EULER	EULER	KUTTA
0.50	0.5000	0.5000	0.5000	0.50	0.5000	0.5000	0.5000
0.60	0.6000	0.6048	0.6049	0.55	0.5500	0.5512	0.5512
0.70	0.7095	0.7191	0.7194	0.60	0.6024	0.6049	0.6049
0.80	0.8283	0.8427	0.8431	0.65	0.6573	0.6609	0.6610
0.90	0.9559	0.9752	0.9757	0.70	0.7144	0.7193	0.7194
1.00	1.0921	1.1163	1.1169	0.75	0.7739	0.7800	0.7801
				0.80	0.8356	0.8430	0.8431
				0.85	0.8996	0.9082	0.9083
				0.90	0.9657	0.9755	0.9757
				0.95	1.0340	1.0451	1.0452
				1.00	1.1044	1.1168	1.1169

6.

$x(n)$	$y(n)$	
0.00	2.0000	initial condition
0.10	2.4734	Runge–Kutta
0.20	3.1781	Runge–Kutta
0.30	4.3925	Runge–Kutta
	6.7689	*predictor*
0.40	7.0783	corrector

10 Plane Autonomous Systems and Stability

_____ **Exercises 10.1** _____

3. The corresponding plane autonomous system is

$$x' = y, \quad y' = x^2 - y(1 - x^3).$$

If (x, y) is a critical point, $y = 0$ and so $x^2 - y(1 - x^3) = x^2 = 0$. Therefore $(0, 0)$ is the sole critical point.

6. The corresponding plane autonomous system is

$$x' = y, \quad y' = -x + \epsilon x|x|.$$

If (x, y) is a critical point, $y = 0$ and $-x + \epsilon x|x| = x(-1 + \epsilon|x|) = 0$. Hence $x = 0$, $1/\epsilon$, $-1/\epsilon$. The critical points are $(0, 0)$, $(1/\epsilon, 0)$ and $(-1/\epsilon, 0)$.

9. From $x - y = 0$ we have $y = x$. Substituting into $3x^2 - 4y = 0$ we obtain $3x^2 - 4x = x(3x - 4) = 0$. It follows that $(0, 0)$ and $(4/3, 4/3)$ are the critical points of the system.

12. Adding the two equations we obtain $10 - 15\dfrac{y}{y + 5} = 0$. It follows that $y = 10$, and from $-2x + y + 10 = 0$ we may conclude that $x = 10$. Therefore $(10, 10)$ is the sole critical point of the system.

15. From $x(1 - x^2 - 3y^2) = 0$ we have $x = 0$ or $x^2 + 3y^2 = 1$. If $x = 0$, then substituting into $y(3 - x^2 - 3y^2)$ gives $y(3 - 3y^2) = 0$. Therefore $y = 0$, 1, -1. Likewise $x^2 = 1 - 3y^2$ yields $2y = 0$ so that $y = 0$ and $x^2 = 1 - 3(0)^2 = 1$. The critical points of the system are therefore $(0, 0)$, $(0, 1)$, $(0, -1)$, $(1, 0)$, and $(-1, 0)$.

18. **(a)** From Exercises 8.2, Problem 6, $x = c_1 + 2c_2 e^{-5t}$ and $y = 3c_1 + c_2 e^{-5t}$.

(b) From $\mathbf{X}(0) = (3, 4)$ it follows that $c_1 = c_2 = 1$. Therefore $x = 1 + 2e^{-5t}$ and $y = 3 + e^{-5t}$.

(c)

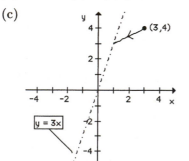

21. (a) From Exercises 8.2, Problem 35, $x = c_1(\sin t - \cos t)e^{4t} + c_2(-\sin t - \cos t)e^{4t}$ and $y = 2c_1(\cos t)e^{4t} + 2c_2(\sin t)e^{4t}$. Because of the presence of e^{4t}, there are no periodic solutions.

(b) From $\mathbf{X}(0) = (-1, 2)$ it follows that $c_1 = 1$ and $c_2 = 0$. Therefore $x = (\sin t - \cos t)e^{4t}$ and $y = 2(\cos t)e^{4t}$.

(c)

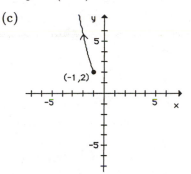

24. Switching to polar coordinates,

$$\frac{dr}{dt} = \frac{1}{r}\left(x\frac{dx}{dt} + y\frac{dy}{dt}\right) = \frac{1}{r}(xy - x^2r^2 - xy + y^2r^2) = r^3$$

$$\frac{d\theta}{dt} = \frac{1}{r^2}\left(-y\frac{dx}{dt} + x\frac{dy}{dt}\right) = \frac{1}{r^2}(-y^2 - xyr^2 - x^2 + xyr^2) = -1.$$

If we use separation of variables, it follows that

$$r = \frac{1}{\sqrt{-2t + c_1}} \quad \text{and} \quad \theta = -t + c_2.$$

Since $\mathbf{X}(0) = (4, 0)$, $r = 4$ and $\theta = 0$ when $t = 0$. It follows that $c_2 = 0$ and $c_1 = \frac{1}{16}$. The final solution may be written as

$$r = \frac{4}{\sqrt{1 - 32t}}, \quad \theta = -t.$$

Note that $r \to \infty$ as $t \to \left(\frac{1}{32}\right)^-$. Because $0 \le t \le \frac{1}{32}$, the curve is not a spiral.

27. The system has no critical points, so there are no periodic solutions.

30. The system has no critical points, so there are no periodic solutions.

———————— **Exercises 10.2** ————————

3. (a) All solutions are unstable spirals which become unbounded as t increases.

(b)

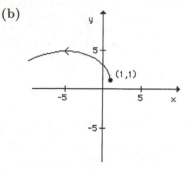

6. (a) All solutions become unbounded and $y = x/2$ serves as the asymptote.

(b)

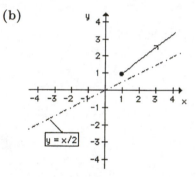

9. Since $\Delta = -41 < 0$, we may conclude from Figure 10.18 that $(0,0)$ is a saddle point.

12. Since $\Delta = 1$ and $\tau = -1$, $\tau^2 - 4\Delta = -3$ and so from Figure 10.18, $(0,0)$ is a stable spiral point.

15. Since $\Delta = 0.01$ and $\tau = -0.03$, $\tau^2 - 4\Delta < 0$ and so from Figure 10.18, $(0,0)$ is a stable spiral point.

18. Note that $\Delta = 1$ and $\tau = \mu$. Therefore we need both $\tau = \mu < 0$ and $\tau^2 - 4\Delta = \mu^2 - 4 < 0$ for $(0,0)$ to be a stable spiral point. These two conditions may be written as $-2 < \mu < 0$.

21. $\mathbf{AX_1 + F = 0}$ implies that $\mathbf{AX_1} = -\mathbf{F}$ or $\mathbf{X_1} = -\mathbf{A}^{-1}\mathbf{F}$. Since $\mathbf{X}_p(t) = -\mathbf{A}^{-1}\mathbf{F}$ is a particular solution, it follows from Theorem 8.6 that $\mathbf{X}(t) = \mathbf{X}_c(t) + \mathbf{X_1}$ is the general solution to $\mathbf{X'} = \mathbf{AX + F}$. If $\tau < 0$ and $\Delta > 0$ then $\mathbf{X}_c(t)$ approaches $(0,0)$ by Theorem 10.1(a). It follows that $\mathbf{X}(t)$ approaches $\mathbf{X_1}$ as $t \to \infty$.

24. (a) The critical point is $\mathbf{X_1} = (-1, -2)$.

(b) From the graph, $\mathbf{X}_1$ appears to be a stable node or a degenerate stable node.

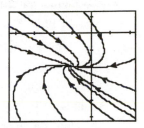

(c) Since $\tau = -16$, $\Delta = 64$, and $\tau^2 - 4\Delta = 0$, $(0,0)$ is a degenerate stable node.

Exercises 10.3

3. The critical points are $x = 0$ and $x = n+1$. Since $g'(x) = k(n+1) - 2kx$, $g'(0) = k(n+1) > 0$ and $g'(n+1) = -k(n+1) < 0$. Therefore $x = 0$ is unstable while $x = n+1$ is asymptotically stable. See Theorem 10.2.

6. The only critical point is $v = mg/k$. Now $g(v) = g - (k/m)v$ and so $g'(v) = -k/m < 0$. Therefore $v = mg/k$ is an asymptotically stable critical point by Theorem 10.2.

9. Critical points occur at $P = a/b$, c but not at $P = 0$. Since $g'(P) = (a - bP) + (P - c)(-b)$,

$$g'(a/b) = (a/b - c)(-b) = -a + bc \quad \text{and} \quad g'(c) = a - bc.$$

Since $a < bc$, $-a + bc > 0$ and $a - bc < 0$. Therefore $P = a/b$ is unstable while $P = c$ is asymptotically stable.

12. Critical points are $(1,0)$ and $(-1,0)$, and

$$\mathbf{g}'(\mathbf{X}) = \begin{pmatrix} 2x & -2y \\ 0 & 2 \end{pmatrix}.$$

At $\mathbf{X} = (1,0)$, $\tau = 4$, $\Delta = 4$, and so $\tau^2 - 4\Delta = 0$. We may conclude that $(1,0)$ is unstable but we are unable to classify this critical point any further. At $\mathbf{X} = (-1,0)$, $\Delta = -4 < 0$ and so $(-1,0)$ is a saddle point.

15. Since $x^2 - y^2 = 0$, $y^2 = x^2$ and so $x^2 - 3x + 2 = (x - 1)(x - 2) = 0$. It follows that the critical points are $(1,1)$, $(1,-1)$, $(2,2)$, and $(2,-2)$. We next use the Jacobian

$$\mathbf{g}'(\mathbf{X}) = \begin{pmatrix} -3 & 2y \\ 2x & -2y \end{pmatrix}$$

to classify these four critical points. For $\mathbf{X} = (1,1)$, $\tau = -5$, $\Delta = 2$, and so $\tau^2 - 4\Delta = 17 > 0$. Therefore $(1,1)$ is a stable node. For $\mathbf{X} = (1,-1)$, $\Delta = -2 < 0$ and so $(1,-1)$ is a saddle point. For $\mathbf{X} = (2,2)$, $\Delta = -4 < 0$ and so we have another saddle point. Finally, if $\mathbf{X} = (2,-2)$, $\tau = 1$, $\Delta = 4$, and so $\tau^2 - 4\Delta = -15 < 0$. Therefore $(2,-2)$ is an unstable spiral point.

18. We found that $(0,0)$, $(0,1)$, $(0,-1)$, $(1,0)$ and $(-1,0)$ were the critical points in Exercise 15, Section 10.1. The Jacobian is

$$\mathbf{g}'(\mathbf{X}) = \begin{pmatrix} 1 - 3x^2 - 3y^2 & -6xy \\ -2xy & 3 - x^2 - 9y^2 \end{pmatrix}.$$

For $\mathbf{X} = (0,0)$, $\tau = 4$, $\Delta = 3$ and so $\tau^2 - 4\Delta = 4 > 0$. Therefore $(0,0)$ is an unstable node. Both $(0,1)$ and $(0,-1)$ give $\tau = -8$, $\Delta = 12$, and $\tau^2 - 4\Delta = 16 > 0$. These two critical points are therefore stable nodes. For $\mathbf{X} = (1,0)$ or $(-1,0)$, $\Delta = -4 < 0$ and so saddle points occur.

21. The corresponding plane autonomous system is

$$\theta' = y, \quad y' = (\cos\theta - \tfrac{1}{2})\sin\theta.$$

Since $|\theta| < \pi$, it follows that critical points are $(0,0)$, $(\pi/3, 0)$ and $(-\pi/3, 0)$. The Jacobian matrix is

$$\mathbf{g}'(\mathbf{X}) = \begin{pmatrix} 0 & 1 \\ \cos 2\theta - \tfrac{1}{2}\cos\theta & 0 \end{pmatrix}$$

and so at $(0,0)$, $\tau = 0$ and $\Delta = -1/2$. Therefore $(0,0)$ is a saddle point. For $\mathbf{X} = (\pm\pi/3, 0)$, $\tau = 0$ and $\Delta = 3/4$. It is not possible to classify either critical point in this borderline case.

24. The corresponding plane autonomous system is

$$x' = y, \quad y' = -\frac{4x}{1+x^2} - 2y$$

and the only critical point is $(0,0)$. Since the Jacobian matrix is

$$\mathbf{g}'(\mathbf{X}) = \begin{pmatrix} 0 & 1 \\ -4\dfrac{1-x^2}{(1+x^2)^2} & -2 \end{pmatrix},$$

$\tau = -2$, $\Delta = 4$, $\tau^2 - 4\Delta = -12$, and so $(0,0)$ is a stable spiral point.

27. The corresponding plane autonomous system is

$$x' = y, \quad y' = -\frac{(\beta + \alpha^2 y^2)x}{1+\alpha^2 x^2}$$

and the Jacobian matrix is

$$\mathbf{g}'(\mathbf{X}) = \begin{pmatrix} 0 & 1 \\ \dfrac{(\beta + \alpha y^2)(\alpha^2 x^2 - 1)}{(1+\alpha^2 x^2)^2} & \dfrac{-2\alpha^2 yx}{1+\alpha^2 x^2} \end{pmatrix}.$$

For $\mathbf{X} = (0,0)$, $\tau = 0$ and $\Delta = \beta$. Since $\beta < 0$, we may conclude that $(0,0)$ is a saddle point.

30. (a) The corresponding plane autonomous system is

$$x' = y, \quad y' = \epsilon(y - \tfrac{1}{3}y^3) - x$$

and so the only critical point is $(0,0)$. Since the Jacobian matrix is

$$\mathbf{g}'(\mathbf{X}) = \begin{pmatrix} 0 & 1 \\ -1 & \epsilon(1-y^2) \end{pmatrix},$$

$\tau = \epsilon$, $\Delta = 1$, and so $\tau^2 - 4\Delta = \epsilon^2 - 4$ at the critical point $(0,0)$.

(b) When $\tau = \epsilon > 0$, $(0,0)$ is an unstable critical point.

(c) When $\epsilon < 0$ and $\tau^2 - 4\Delta = \epsilon^2 - 4 < 0$, $(0,0)$ is a stable spiral point. These two requirements can be written as $-2 < \epsilon < 0$.

(d) When $\epsilon = 0$, $x'' + x = 0$ and so $x = c_1 \cos t + c_2 \sin t$. Therefore all solutions are periodic (with period 2π) and so $(0,0)$ is a center.

33. (a) $x' = 2xy = 0$ implies that either $x = 0$ or $y = 0$. If $x = 0$, then from $1 - x^2 + y^2 = 0$, $y^2 = -1$ and there are no real solutions. If $y = 0$, $1 - x^2 = 0$ and so $(1,0)$ and $(-1,0)$ are critical points. The Jacobian matrix is

$$\mathbf{g}'(\mathbf{X}) = \begin{pmatrix} 2y & 2x \\ -2x & 2y \end{pmatrix}$$

and so $\tau = 0$ and $\Delta = 4$ at either $\mathbf{X} = (1,0)$ or $(-1,0)$. We obtain no information about these critical points in this borderline case.

(b) $\dfrac{dy}{dx} = \dfrac{y'}{x'} = \dfrac{1 - x^2 + y^2}{2xy}$ or $2xy\dfrac{dy}{dx} = 1 - x^2 + y^2$. Letting $\mu = \dfrac{y^2}{x}$, it follows that $\dfrac{d\mu}{dx} = \dfrac{1}{x^2} - 1$ and so $\mu = -\dfrac{1}{x} - x + 2c$. Therefore $\dfrac{y^2}{x} = -\dfrac{1}{x} - x + 2c$ which can be put in the form

$$(x - c)^2 + y^2 = c^2 - 1.$$

The solution curves are shown and so both $(1,0)$ and $(-1,0)$ are centers.

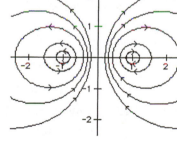

36. The corresponding plane autonomous system is

$$x' = y, \quad y' = \epsilon x^2 - x + 1$$

and so the critical points must satisfy $y = 0$ and

$$x = \frac{1 \pm \sqrt{1 - 4\epsilon}}{2\epsilon}.$$

Therefore we must require that $\epsilon \le \frac{1}{4}$ for real solutions to exist. We will use the Jacobian matrix

$$\mathbf{g}'(\mathbf{X}) = \begin{pmatrix} 0 & 1 \\ 2\epsilon x - 1 & 0 \end{pmatrix}$$

to attempt to classify $((1 \pm \sqrt{1-4\epsilon})/2\epsilon, 0)$ when $\epsilon \le 1/4$. Note that $\tau = 0$ and $\Delta = \mp\sqrt{1-4\epsilon}$. For $\mathbf{X} = ((1 + \sqrt{1-4\epsilon})/2\epsilon, 0)$ and $\epsilon < 1/4$, $\Delta < 0$ and so a saddle point occurs. For $\mathbf{X} = ((1 - \sqrt{1-4\epsilon})/2\epsilon, 0)$, $\Delta \ge 0$ and we are not able to classify this critical point using linearization.

39. (a) Letting $x = \theta$ and $y = x'$ we obtain the system $x' = y$ and $y' = 1/2 - \sin x$. Since $\sin \pi/6 = \sin 5\pi/6 = 1/2$ we see that $(\pi/6, 0)$ and $(5\pi/6, 0)$ are critical points of the system.

(b) The Jacobian matrix is

$$\mathbf{g}'(\mathbf{X}) = \begin{pmatrix} 0 & 1 \\ -\cos x & 0 \end{pmatrix}$$

and so

$$\mathbf{A}_1 = \mathbf{g}' = ((\pi/6, 0)) = \begin{pmatrix} 0 & 1 \\ -\sqrt{3}/2 & 0 \end{pmatrix} \quad \text{and} \quad \mathbf{A}_2 = \mathbf{g}' = ((5\pi/6, 0)) = \begin{pmatrix} 0 & 1 \\ \sqrt{3}/2 & 0 \end{pmatrix}.$$

Since $\det \mathbf{A}_1 > 0$ and the trace of $\mathbf{A}_1$ is 0, no conclusion can be drawn regarding the critical point $(\pi/6, 0)$. Since $\det \mathbf{A}_2 < 0$, we see that $(5\pi/6, 0)$ is a saddle point.

(c) From the system in part (a) we obtain the first-order differential equation

$$\frac{dy}{dx} = \frac{1/2 - \sin x}{y}.$$

Separating variables and integrating we obtain

$$\int y \, dy = \int \left(\frac{1}{2} - \sin x\right) dx$$

and

$$\frac{1}{2}y^2 = \frac{1}{2}x + \cos x + c_1$$

or

$$y^2 = x + 2\cos x + c_2.$$

For x_0 near $\pi/6$, if $\mathbf{X}(0) = (x_0, 0)$ then $c_2 = -x_0 - 2\cos x_0$ and $y^2 = x + 2\cos x - x_0 - 2\cos x_0$. Thus, there are two values of y for each x in a sufficiently small interval around $\pi/6$. Therefore $(\pi/6, 0)$ is a center.

Exercises 10.4

3. The corresponding plane autonomous system is

$$x' = y, \quad y' = -g\frac{f'(x)}{1 + [f'(x)]^2} - \frac{\beta}{m}y$$

and

$$\frac{\partial}{\partial x}\left(-g\frac{f'(x)}{1 + [f'(x)]^2} - \frac{\beta}{m}y\right) = -g\frac{(1 + [f'(x)]^2)f''(x) - f'(x)2f'(x)f''(x)}{(1 + [f'(x)]^2)^2}.$$

If $X_1 = (x_1, y_1)$ is a critical point, $y_1 = 0$ and $f'(x_1) = 0$. The Jacobian at this critical point is therefore

$$g'(X_1) = \begin{pmatrix} 0 & 1 \\ -gf''(x_1) & -\frac{\beta}{m} \end{pmatrix}.$$

6. (a) If $f(x) = \cosh x$, $f'(x) = \sinh x$ and $[f'(x)]^2 + 1 = \sinh^2 x + 1 = \cosh^2 x$. Therefore

$$\frac{dy}{dx} = \frac{y'}{x'} = -g\frac{\sinh x}{\cosh^2 x}\frac{1}{y}.$$

We may separate variables to show that $y^2 = \dfrac{2g}{\cosh x} + c$. But $x(0) = x_0$ and $y(0) = x'(0) = v_0$.

Therefore $c = v_0^2 - \dfrac{2g}{\cosh x_0}$ and so

$$y^2 = \frac{2g}{\cosh x} - \frac{2g}{\cosh x_0} + v_0^2.$$

Now

$$\frac{2g}{\cosh x} - \frac{2g}{\cosh x_0} + v_0^2 \geq 0 \quad \text{if and only if} \quad \cosh x \leq \frac{2g \cosh x_0}{2g - v_0^2 \cosh x_0}$$

and the solution to this inequality is an interval $[-a, a]$. Therefore each x in $(-a, a)$ has two corresponding values of y and so the solution is periodic.

(b) Since $z = \cosh x$, the maximum height occurs at the largest value of x on the cycle. From (a), $x_{\max} = a$ where $\cosh a = \dfrac{2g \cosh x_0}{2g - v_0^2 \cosh x_0}$. Therefore

$$z_{\max} = \frac{2g \cosh x_0}{2g - v_0^2 \cosh x_0}.$$

9. (a) In the Lotka-Volterra Model the average number of predators is d/c and the average number of prey is a/b. But

$$x' = -ax + bxy - \epsilon_1 x = -(a + \epsilon_1)x + bxy$$

$$y' = -cxy + dy - \epsilon_2 y = -cxy + (d - \epsilon_2)y$$

and so the new critical point in the first quadrant is $(d/c - \epsilon_2/c, a/b + \epsilon_1/b)$.

(b) The average number of predators $d/c - \epsilon_2/c$ has decreased while the average number of prey $a/b + \epsilon_1/b$ has increased. The fishery science model is consistent with Volterra's principle.

12. $\Delta = r_1 r_2$, $\tau = r_1 + r_2$ and $\tau^2 - 4\Delta = (r_1 + r_2)^2 - 4r_1 r_2 = (r_1 - r_2)^2$. Therefore when $r_1 \neq r_2$, $(0,0)$ is an unstable node.

15. $\dfrac{K_1}{\alpha_{12}} < K_2 < K_1 \alpha_{21}$ and so $\alpha_{12}\alpha_{21} > 1$. Therefore $\Delta = (1 - \alpha_{12}\alpha_{21})\hat{x}\hat{y}\dfrac{r_1 r_2}{K_1 K_2} < 0$ and so $(\hat{x}, \hat{y})$ is a saddle point.

Chapter 10 Review Exercises

3. A center or a saddle point

6. True

9. True

12. (a) If $\mathbf{X}(0) = \mathbf{X}_0$ lies on the line $y = -2x$, then $\mathbf{X}(t)$ approaches $(0,0)$ along this line. For all other initial conditions, $\mathbf{X}(t)$ approaches $(0,0)$ from the direction determined by the line $y = x$.

    (b) If $\mathbf{X}(0) = \mathbf{X}_0$ lies on the line $y = -x$, then $\mathbf{X}(t)$ approaches $(0,0)$ along this line. For all other initial conditions, $\mathbf{X}(t)$ becomes unbounded and $y = 2x$ serves as an asymptote.

15. From $x = r\cos\theta$, $y = r\sin\theta$ we have

$$\frac{dx}{dt} = -r\sin\theta\,\frac{d\theta}{dt} + \frac{dr}{dt}\cos\theta$$

$$\frac{dy}{dt} = r\cos\theta\,\frac{d\theta}{dt} + \frac{dr}{dt}\sin\theta.$$

Then $r' = \alpha r$, $\theta' = 1$ gives

$$\frac{dx}{dt} = -r\sin\theta + \alpha r\cos\theta$$

$$\frac{dy}{dt} = r\cos\theta + \alpha r\sin\theta.$$

We see that $r = 0$, which corresponds to $\mathbf{X} = (0,0)$, is a critical point. Solving $r' = \alpha r$ we have $r = c_1 e^{\alpha t}$. Thus, when $\alpha < 0$, $\lim_{t\to\infty} r(t) = 0$ and $(0,0)$ is a stable critical point. When $\alpha = 0$, $r' = 0$ and $r = c_1$. In this case $(0,0)$ is a center, which is stable. Therefore, $(0,0)$ is a stable critical point for the system when $\alpha \leq 0$.

18. $\dfrac{dy}{dx} = \dfrac{y'}{x'} = \dfrac{-2x\sqrt{y^2+1}}{y}$. We may separate variables to show that $\sqrt{y^2+1} = -x^2 + c$. But $x(0) = x_0$ and $y(0) = x'(0) = 0$. It follows that $c = 1 + x_0^2$ so that

$$y^2 = (1 + x_0^2 - x^2)^2 - 1.$$

Note that $1 + x_0^2 - x^2 > 1$ for $-x_0 < x < x_0$ and $y = 0$ for $x = \pm x_0$. Each x with $-x_0 < x < x_0$ has two corresponding values of y and so the solution $\mathbf{X}(t)$ with $\mathbf{X}(0) = (x_0, 0)$ is periodic.

11 Orthogonal Functions and Fourier Series

Exercises 11.1

3. $\int_0^2 e^x (xe^{-x} - e^{-x}) dx = \int_0^2 (x-1)dx = \left(\frac{1}{2}x^2 - x \right) \Big|_0^2 = 0$

6. $\int_{\pi/4}^{5\pi/4} e^x \sin x \, dx = \left(\frac{1}{2}e^x \sin x - \frac{1}{2}e^x \cos x \right) \Big|_{\pi/4}^{5\pi/4} = 0$

9. For $m \neq n$

$$\int_0^{\pi} \sin nx \sin mx \, dx = \frac{1}{2} \int_0^{\pi} [\cos(n-m)x - \cos(n+m)x] \, dx$$

$$= \frac{1}{2(n-m)} \sin(n-m)x \Big|_0^{\pi} - \frac{1}{2(n+m)} \sin 2(n+m)x \Big|_0^{\pi}$$

$$= 0.$$

For $m = n$

$$\int_0^{\pi} \sin^2 nx \, dx = \int_0^{\pi} \left[\frac{1}{2} - \frac{1}{2}\cos 2nx \right] dx = \frac{1}{2}x \Big|_0^{\pi} - \frac{1}{4n} \sin 2nx \Big|_0^{\pi} = \frac{\pi}{2}$$

so that

$$\| \sin nx \| = \sqrt{\frac{\pi}{2}}.$$

12. For $m \neq n$, we have from Problems 11 and 10 that

$$\int_{-p}^{p} \cos \frac{n\pi}{p}x \cos \frac{m\pi}{p}x \, dx = 2 \int_0^{p} \cos \frac{n\pi}{p}x \cos \frac{m\pi}{p}x \, dx = 0$$

$$\int_{-p}^{p} \sin \frac{n\pi}{p}x \sin \frac{m\pi}{p}x \, dx = 2 \int_0^{p} \sin \frac{n\pi}{p}x \sin \frac{m\pi}{p}x \, dx = 0.$$

Also

$$\int_{-p}^{p} \sin \frac{n\pi}{p}x \cos \frac{m\pi}{p}x \, dx = \frac{1}{2} \int_{-p}^{p} \left(\sin \frac{(n-m)\pi}{p}x + \sin \frac{(n+m)\pi}{p}x \right) dx = 0,$$

$$\int_{-p}^{p} 1 \cdot \cos \frac{n\pi}{p}x \, dx = \frac{p}{n\pi} \sin \frac{n\pi}{p}x \Big|_{-p}^{p} = 0,$$

$$\int_{-p}^{p} 1 \cdot \sin \frac{n\pi}{p}x \, dx = -\frac{p}{n\pi} \cos \frac{n\pi}{p}x \Big|_{-p}^{p} = 0,$$

and

$$\int_{-p}^{p} \sin\frac{n\pi}{p}x \cos\frac{n\pi}{p}x\, dx = \int_{-p}^{p} \frac{1}{2}\sin\frac{2n\pi}{p}x\, dx = -\frac{p}{4n\pi}\cos\frac{2n\pi}{p}x \bigg|_{-p}^{p} = 0.$$

For $m = n$

$$\int_{-p}^{p} \cos^2\frac{n\pi}{p}x\, dx = \int_{-p}^{p}\left(\frac{1}{2} + \frac{1}{2}\cos\frac{2n\pi}{p}x\right)dx = p,$$

$$\int_{-p}^{p} \sin^2\frac{n\pi}{p}x\, dx = \int_{-p}^{p}\left(\frac{1}{2} - \frac{1}{2}\cos\frac{2n\pi}{p}x\right)dx = p,$$

and

$$\int_{-p}^{p} 1^2 dx = 2p$$

so that

$$\|1\| = \sqrt{2p}, \quad \left\|\cos\frac{n\pi}{p}x\right\| = \sqrt{p}, \quad \text{and} \quad \left\|\sin\frac{n\pi}{p}x\right\| = \sqrt{p}.$$

15. By orthogonality $\int_a^b \phi_0(x)\phi_n(x)dx = 0$ for $n = 1, 2, 3, \ldots$; that is, $\int_a^b \phi_n(x)dx = 0$ for $n = 1, 2, 3, \ldots$.

18. Setting

$$0 = \int_{-2}^{2} f_3(x)f_1(x)\, dx = \int_{-2}^{2}\left(x^2 + c_1 x^3 + c_2 x^4\right)dx = \frac{16}{3} + \frac{64}{5}c_2$$

and

$$0 = \int_{-2}^{2} f_3(x)f_2(x)\, dx = \int_{-2}^{2}\left(x^3 + c_1 x^4 + c_2 x^5\right)dx = \frac{64}{5}c_1$$

we obtain $c_1 = 0$ and $c_2 = -5/12$.

Exercises 11.2

3. $a_0 = \int_{-1}^{1} f(x)\, dx = \int_{-1}^{0} 1\, dx + \int_{0}^{1} x\, dx = \frac{3}{2}$

$a_n = \int_{-1}^{1} f(x)\cos n\pi x\, dx = \int_{-1}^{0} \cos n\pi x\, dx + \int_{0}^{1} x\cos n\pi x\, dx = \frac{1}{n^2\pi^2}[(-1)^n - 1]$

$b_n = \int_{-1}^{1} f(x)\sin n\pi x\, dx = \int_{-1}^{0} \sin n\pi x\, dx + \int_{0}^{1} x\sin n\pi x\, dx = -\frac{1}{n\pi}$

$f(x) = \frac{3}{4} + \sum_{n=1}^{\infty}\left[\frac{(-1)^n - 1}{n^2\pi^2}\cos n\pi x - \frac{1}{n\pi}\sin n\pi x\right]$

6. $a_0 = \frac{1}{\pi}\int_{-\pi}^{\pi} f(x)\, dx = \frac{1}{\pi}\int_{-\pi}^{0} \pi^2\, dx + \frac{1}{\pi}\int_{0}^{\pi}\left(\pi^2 - x^2\right)dx = \frac{5}{3}\pi^2$

$$a_n = \frac{1}{\pi} \int_{-\pi}^{\pi} f(x) \cos nx \, dx = \frac{1}{\pi} \int_{-\pi}^{0} \pi^2 \cos nx \, dx + \frac{1}{\pi} \int_{0}^{\pi} \left(\pi^2 - x^2\right) \cos nx \, dx$$

$$= \frac{1}{\pi} \left(\frac{\pi^2 - x^2}{n} \sin nx \, \Big|_0^{\pi} + \frac{2}{n} \int_0^{\pi} x \sin nx \, dx \right) = \frac{2}{n^2}(-1)^{n+1}$$

$$b_n = \frac{1}{\pi} \int_{-\pi}^{\pi} f(x) \sin nx \, dx = \frac{1}{\pi} \int_{-\pi}^{0} \pi^2 \sin nx \, dx + \frac{1}{\pi} \int_{0}^{\pi} \left(\pi^2 - x^2\right) \sin nx \, dx$$

$$= \frac{\pi}{n}[(-1)^n - 1] + \frac{1}{\pi}\left(\frac{x^2 - \pi^2}{n}\cos nx \,\Big|_0^{\pi} - \frac{2}{n}\int_0^{\pi} x \cos nx \, dx\right) = \frac{\pi}{n}(-1)^n + \frac{2}{n^3\pi}[1 - (-1)^n]$$

$$f(x) = \frac{5\pi^2}{6} + \sum_{n=1}^{\infty} \left[\frac{2}{n^2}(-1)^{n+1} \cos nx + \left(\frac{\pi}{n}(-1)^n + \frac{2[1 - (-1)^n]}{n^3\pi} \right) \sin nx \right]$$

9. $a_0 = \dfrac{1}{\pi} \displaystyle\int_{-\pi}^{\pi} f(x)\, dx = \dfrac{1}{\pi} \displaystyle\int_{0}^{\pi} \sin x \, dx = \dfrac{2}{\pi}$

$$a_n = \frac{1}{\pi} \int_{-\pi}^{\pi} f(x) \cos nx \, dx = \frac{1}{\pi} \int_{0}^{\pi} \sin x \cos nx \, dx = \frac{1}{2\pi} \int_{0}^{\pi} [\sin(n+1)x + \sin(1-n)x] \, dx$$

$$= \frac{1 + (-1)^n}{\pi(1 - n^2)} \quad \text{for } n = 2, 3, 4, \ldots$$

$$a_1 = \frac{1}{2\pi} \int_0^{\pi} \sin 2x \, dx = 0$$

$$b_n = \frac{1}{\pi} \int_{-\pi}^{\pi} f(x) \sin nx \, dx = \frac{1}{\pi} \int_0^{\pi} \sin x \sin nx \, dx$$

$$= \frac{1}{2\pi} \int_0^{\pi} [\cos(1-n)x - \cos(1+n)x] \, dx = 0 \quad \text{for } n = 2, 3, 4, \ldots$$

$$b_1 = \frac{1}{2\pi} \int_0^{\pi} (1 - \cos 2x) \, dx = \frac{1}{2}$$

$$f(x) = \frac{1}{\pi} + \frac{1}{2} \sin x + \sum_{n=2}^{\infty} \frac{1 + (-1)^n}{\pi(1 - n^2)} \cos nx$$

12. $a_0 = \dfrac{1}{2} \displaystyle\int_{-2}^{2} f(x)\, dx = \dfrac{1}{2}\left(\displaystyle\int_0^1 x \, dx + \displaystyle\int_1^2 1 \, dx \right) = \dfrac{3}{4}$

$$a_n = \frac{1}{2} \int_{-2}^{2} f(x) \cos \frac{n\pi}{2} x \, dx = \frac{1}{2}\left(\int_0^1 x \cos \frac{n\pi}{2} x \, dx + \int_1^2 \cos \frac{n\pi}{2} x \, dx \right) = \frac{2}{n^2\pi^2}\left(\cos \frac{n\pi}{2} - 1 \right)$$

$$b_n = \frac{1}{2} \int_{-2}^{2} f(x) \sin \frac{n\pi}{2} x \, dx = \frac{1}{2}\left(\int_0^1 x \sin \frac{n\pi}{2} x \, dx + \int_1^2 \sin \frac{n\pi}{2} x \, dx \right)$$

$$= \frac{2}{n^2\pi^2}\left(\sin \frac{n\pi}{2} + \frac{n\pi}{2}(-1)^{n+1} \right)$$

$$f(x) = \frac{3}{8} + \sum_{n=1}^{\infty} \left[\frac{2}{n^2\pi^2} \left(\cos \frac{n\pi}{2} - 1 \right) \cos \frac{n\pi}{2}x + \frac{2}{n^2\pi^2} \left(\sin \frac{n\pi}{2} + \frac{n\pi}{2}(-1)^{n+1} \right) \sin \frac{n\pi}{2}x \right]$$

15. $a_0 = \dfrac{1}{\pi} \displaystyle\int_{-\pi}^{\pi} f(x)\,dx = \dfrac{1}{\pi} \displaystyle\int_{-\pi}^{\pi} e^x\,dx = \dfrac{1}{\pi}(e^{\pi} - e^{-\pi})$

$a_n = \dfrac{1}{\pi} \displaystyle\int_{-\pi}^{\pi} f(x)\cos nx\,dx = \dfrac{(-1)^n(e^{\pi} - e^{-\pi})}{\pi(1 + n^2)}$

$b_n = \dfrac{1}{\pi} \displaystyle\int_{-\pi}^{\pi} f(x)\sin nx\,dx = \dfrac{1}{\pi} \displaystyle\int_{-\pi}^{\pi} e^x \sin nx\,dx = \dfrac{(-1)^n n(e^{-\pi} - e^{\pi})}{\pi(1 + n^2)}$

$f(x) = \dfrac{e^{\pi} - e^{-\pi}}{2\pi} + \displaystyle\sum_{n=1}^{\infty} \left[\dfrac{(-1)^n(e^{\pi} - e^{-\pi})}{\pi(1 + n^2)} \cos nx + \dfrac{(-1)^n n(e^{-\pi} - e^{\pi})}{\pi(1 + n^2)} \sin nx \right]$

18. From Problem 17

$$\frac{\pi^2}{8} = \frac{1}{2}\left(\frac{\pi^2}{6} + \frac{\pi^2}{12} \right) = \frac{1}{2}\left(2 + \frac{2}{3^2} + \frac{2}{5^2} + \cdots \right) = 1 + \frac{1}{3^2} + \frac{1}{5^2} + \cdots.$$

21. (a) Letting $c_0 = a_0/2$, $c_n = (a_n - ib_n)$, and $c_{-n} = (a_n + ib_n)/2$ we have

$$f(x) = \frac{a_0}{2} + \sum_{n=1}^{\infty}\left(a_n \cos \frac{n\pi}{p}x + b_n \sin \frac{n\pi}{p}x \right)$$

$$= c_0 + \sum_{n=1}^{\infty}\left(a_n \frac{e^{in\pi x/p} + e^{-in\pi x/p}}{2} + b_n \frac{e^{in\pi x/p} - e^{-in\pi x/p}}{2i} \right)$$

$$= c_0 + \sum_{n=1}^{\infty}\left(a_n \frac{e^{in\pi x/p} + e^{-in\pi x/p}}{2} - b_n \frac{ie^{in\pi x/p} - ie^{-in\pi x/p}}{2} \right)$$

$$= c_0 + \sum_{n=1}^{\infty}\left(\frac{a_n - ib_n}{2} e^{in\pi x/p} + \frac{a_n + ib_n}{2} e^{-in\pi x/p} \right)$$

$$= c_0 + \sum_{n=1}^{\infty}\left(c_n e^{in\pi x/p} + c_{-n} e^{i(-n)\pi x/p} \right) = \sum_{n=-\infty}^{\infty} c_n e^{in\pi x/p}.$$

(b) Multiplying both sides of the expression in (a) by $e^{-im\pi x/p}$ and integrating we obtain

$$\int_{-p}^{p} f(x)e^{-im\pi x/p}dx = \int_{-p}^{p}\left(\sum_{n=-\infty}^{\infty} c_n e^{in\pi x/p} e^{-im\pi x/p} \right) dx$$

$$= \sum_{n=-\infty}^{\infty} c_n \int_{-p}^{p} e^{i(n-m)\pi x/p}dx$$

$$= \sum_{n\neq m} c_n \int_{-p}^{p} e^{i(n-m)\pi x/p}dx + c_m \int_{-p}^{p} e^{i(m-n)\pi x/p}dx$$

$$= \sum_{n \neq m} c_n \int_{-p}^{p} e^{i(n-m)\pi x/p} dx + c_m \int_{-p}^{p} dx$$

$$= \sum_{n \neq m} c_n \int_{-p}^{p} e^{i(n-m)\pi x/p} dx + 2pc_m.$$

Recalling that $e^{iy} = \cos y + i \sin y$ and $e^{-iy} = \cos y - i \sin y$, we have for $n - m$ an integer and $n \neq m$

$$\int_{-p}^{p} e^{i(n-m)\pi x/p} dx = \frac{p}{i(n-m)\pi} e^{i(n-m)\pi x/p} \Big|_{-p}^{p}$$

$$= \frac{p}{i(n-m)\pi} \left(e^{i(n-m)\pi} - e^{-i(n-m)\pi} \right)$$

$$= \frac{p}{i(n-m)\pi} [\cos(n-m)\pi + i \sin(n-m)\pi - \cos(n-m)\pi + i \sin(n-m)\pi]$$

$$= 0.$$

Thus

$$\int_{-p}^{p} f(x) e^{-im\pi x/p} dx = 2pc_m$$

and

$$c_m = \frac{1}{2p} \int_{-p}^{p} f(x) e^{-im\pi x/p} dx.$$

Exercises 11.3

3. Since $f(-x) = (-x)^2 - x = x^2 - x$, $f(x)$ is neither even nor odd.

6. Since $f(-x) = e^{-x} - e^{x} = -f(x)$, $f(x)$ is an odd function.

9. Since $f(x)$ is not defined for $x < 0$, it is neither even nor odd.

12. Since $f(x)$ is an even function, we expand in a cosine series:

$$a_0 = \int_{1}^{2} 1 \, dx = 1$$

$$a_n = \int_{1}^{2} \cos \frac{n\pi}{2} x \, dx = -\frac{2}{n\pi} \sin \frac{n\pi}{2}.$$

Thus

$$f(x) = \frac{1}{2} + \sum_{n=1}^{\infty} \frac{-2}{n\pi} \sin \frac{n\pi}{2} \cos \frac{n\pi}{2} x.$$

15. Since $f(x)$ is an even function, we expand in a cosine series:

$$a_0 = 2 \int_0^1 x^2 \, dx = \frac{2}{3}$$

$$a_n = 2 \int_0^1 x^2 \cos n\pi x \, dx = 2 \left(\frac{x^2}{n\pi} \sin n\pi x \Big|_0^1 - \frac{2}{n\pi} \int_0^1 x \sin n\pi x \, dx \right) = \frac{4}{n^2\pi^2}(-1)^n.$$

Thus

$$f(x) = \frac{1}{3} + \sum_{n=1}^{\infty} \frac{4}{n^2\pi^2}(-1)^n \cos n\pi x.$$

18. Since $f(x)$ is an odd function, we expand in a sine series:

$$b_n = \frac{2}{\pi} \int_0^\pi x^3 \sin nx \, dx = \frac{2}{\pi} \left(-\frac{x^3}{n} \cos nx \Big|_0^\pi + \frac{3}{n} \int_0^\pi x^2 \cos nx \, dx \right)$$

$$= \frac{2\pi^2}{n}(-1)^{n+1} - \frac{12}{n^2\pi} \int_0^\pi x \sin nx \, dx$$

$$= \frac{2\pi^2}{n}(-1)^{n+1} - \frac{12}{n^2\pi} \left(-\frac{x}{n} \cos nx \Big|_0^\pi + \frac{1}{n} \int_0^\pi \cos nx \, dx \right) = \frac{2\pi^2}{n}(-1)^{n+1} + \frac{12}{n^3}(-1)^n.$$

Thus

$$f(x) = \sum_{n=1}^{\infty} \left(\frac{2\pi^2}{n}(-1)^{n+1} + \frac{12}{n^3}(-1)^n \right) \sin nx.$$

21. Since $f(x)$ is an even function, we expand in a cosine series:

$$a_0 = \int_0^1 x \, dx + \int_1^2 1 \, dx = \frac{3}{2}$$

$$a_n = \int_0^1 x \cos \frac{n\pi}{2} x \, dx + \int_1^2 \cos \frac{n\pi}{2} x \, dx = \frac{4}{n^2\pi^2} \left(\cos \frac{n\pi}{2} - 1 \right).$$

Thus

$$f(x) = \frac{3}{4} + \sum_{n=1}^{\infty} \frac{4}{n^2\pi^2} \left(\cos \frac{n\pi}{2} - 1 \right) \cos \frac{n\pi}{2} x.$$

24. Since $f(x)$ is an even function, we expand in a cosine series.

$$a_0 = \frac{2}{\pi/2} \int_0^{\pi/2} \cos x \, dx = \frac{4}{\pi}$$

$$a_n = \frac{2}{\pi/2} \int_0^{\pi/2} \cos x \cos \frac{n\pi}{\pi/2} x \, dx = \frac{4}{\pi} \int_0^{\pi/2} \cos x \cos 2nx \, dx$$

$$= \frac{2}{\pi} \int_0^{\pi/2} [\cos(2n-1)x + \cos(2n+1)x] = \frac{4(-1)^{n+1}}{\pi(4n^2-1)}.$$

Thus

$$f(x) = \frac{2}{\pi} + \sum_{n=1}^{\infty} \frac{4(-1)^{n+1}}{\pi\,(4n^2 - 1)}\cos 2nx.$$

27. $a_0 = \dfrac{4}{\pi}\displaystyle\int_0^{\pi/2}\cos x\,dx = \dfrac{4}{\pi}$

$a_n = \dfrac{4}{\pi}\displaystyle\int_0^{\pi/2}\cos x\,\cos 2nx\,dx = \dfrac{2}{\pi}\int_0^{\pi/2}[\cos(2n+1)x + \cos(2n-1)x]\,dx = \dfrac{4(-1)^n}{\pi(1-4n^2)}$

$b_n = \dfrac{4}{\pi}\displaystyle\int_0^{\pi/2}\cos x\,\sin 2nx\,dx = \dfrac{2}{\pi}\int_0^{\pi/2}[\sin(2n+1)x + \sin(2n-1)x]\,dx = \dfrac{8n}{\pi(4n^2-1)}$

$f(x) = \dfrac{2}{\pi} + \displaystyle\sum_{n=1}^{\infty}\dfrac{4(-1)^n}{\pi(1-4n^2)}\cos 2nx$

$f(x) = \displaystyle\sum_{n=1}^{\infty}\dfrac{8n}{\pi(4n^2-1)}\sin 2nx$

30. $a_0 = \dfrac{1}{\pi}\displaystyle\int_\pi^{2\pi}(x-\pi)\,dx = \dfrac{\pi}{2}$

$a_n = \dfrac{1}{\pi}\displaystyle\int_\pi^{2\pi}(x-\pi)\cos\dfrac{n}{2}x\,dx = \dfrac{4}{n^2\pi}\left[(-1)^n - \cos\dfrac{n\pi}{2}\right]$

$b_n = \dfrac{1}{\pi}\displaystyle\int_\pi^{2\pi}(x-\pi)\sin\dfrac{n}{2}x\,dx = \dfrac{2}{n}(-1)^{n+1} - \dfrac{4}{n^2\pi}\sin\dfrac{n\pi}{2}$

$f(x) = \dfrac{\pi}{4} + \displaystyle\sum_{n=1}^{\infty}\dfrac{4}{n^2\pi}\left[(-1)^n - \cos\dfrac{n\pi}{2}\right]\cos\dfrac{n}{2}x$

$f(x) = \displaystyle\sum_{n=1}^{\infty}\left(\dfrac{2}{n}(-1)^{n+1} - \dfrac{4}{n^2\pi}\sin\dfrac{n\pi}{2}\right)\sin\dfrac{n}{2}x$

33. $a_0 = 2\displaystyle\int_0^1(x^2+x)\,dx = \dfrac{5}{3}$

$a_n = 2\displaystyle\int_0^1(x^2+x)\cos n\pi x\,dx = \dfrac{2(x^2+x)}{n\pi}\sin n\pi x\,\Big|_0^1 - \dfrac{2}{n\pi}\int_0^1(2x+1)\sin n\pi x\,dx = \dfrac{2}{n^2\pi^2}[3(-1)^n - 1]$

$b_n = 2\displaystyle\int_0^1(x^2+x)\sin n\pi x\,dx = -\dfrac{2(x^2+x)}{n\pi}\cos n\pi x\,\Big|_0^1 + \dfrac{2}{n\pi}\int_0^1(2x+1)\cos n\pi x\,dx$

$\qquad = \dfrac{4}{n\pi}(-1)^{n+1} + \dfrac{4}{n^3\pi^3}[(-1)^n - 1]$

$f(x) = \dfrac{5}{6} + \displaystyle\sum_{n=1}^{\infty}\dfrac{2}{n^2\pi^2}[3(-1)^n - 1]\cos n\pi x$

$f(x) = \displaystyle\sum_{n=1}^{\infty}\left(\dfrac{4}{n\pi}(-1)^{n+1} + \dfrac{4}{n^3\pi^3}[(-1)^n - 1]\right)\sin n\pi x$

36. $a_0 = \dfrac{2}{\pi}\displaystyle\int_0^\pi x\,dx = \pi$

$a_n = \dfrac{2}{\pi}\displaystyle\int_0^\pi x\cos 2nx\,dx = 0$

$b_n = \dfrac{2}{\pi}\displaystyle\int_0^\pi x\sin 2nx\,dx = -\dfrac{1}{n}$

$f(x) = \dfrac{\pi}{2} + \displaystyle\sum_{n=1}^\infty \left(-\dfrac{1}{n}\sin 2nx\right)$

39. We have

$$b_n = \frac{2}{\pi}\int_0^\pi 5\sin nt\,dt = \frac{10}{n\pi}[1-(-1)^n]$$

so that

$$f(t) = \sum_{n=1}^\infty \frac{10[1-(-1)^n]}{n\pi}\sin nt.$$

Substituting the assumption $x_p(t) = \displaystyle\sum_{n=1}^\infty B_n\sin nt$ into the differential equation then gives

$$x_p'' + 10x_p = \sum_{n=1}^\infty B_n(10-n^2)\sin nt = \sum_{n=1}^\infty \frac{10[1-(-1)^n]}{n\pi}\sin nt$$

and so $B_n = \dfrac{10[1-(-1)^n]}{n\pi(10-n^2)}$. Thus

$$x_p(t) = \frac{10}{\pi}\sum_{n=1}^\infty \frac{1-(-1)^n}{n(10-n^2)}\sin nt.$$

42. We have

$$a_0 = \frac{2}{(1/2)}\int_0^{1/2} t\,dt = \frac{1}{2}$$

$$a_n = \frac{2}{(1/2)}\int_0^{1/2} t\cos 2n\pi t\,dt = \frac{1}{n^2\pi^2}[(-1)^n - 1]$$

so that

$$f(t) = \frac{1}{4} + \sum_{n=1}^\infty \frac{(-1)^n - 1}{n^2\pi^2}\cos 2n\pi t.$$

Substituting the assumption

$$x_p(t) = \frac{A_0}{2} + \sum_{n=1}^\infty A_n\cos 2n\pi t$$

into the differential equation then gives

$$\frac{1}{4}x_p'' + 12x_p = 6A_0 + \sum_{n=1}^\infty A_n(12-n^2\pi^2)\cos 2n\pi t = \frac{1}{4} + \sum_{n=1}^\infty \frac{(-1)^n - 1}{n^2\pi^2}\cos 2n\pi t$$

and $A_0 = \dfrac{1}{24}$, $A_n = \dfrac{(-1)^n - 1}{n^2\pi^2(12 - n^2\pi^2)}$. Thus

$$x_p(t) = \frac{1}{48} + \frac{1}{\pi^2}\sum_{n=1}^{\infty}\frac{(-1)^n - 1}{n^2(12 - n^2\pi^2)}\cos 2n\pi t.$$

45. (a) We have

$$b_n = \frac{2}{L}\int_0^L \frac{w_0 x}{L}\sin\frac{n\pi}{L}x\,dx = \frac{2w_0}{n\pi}(-1)^{n+1}$$

so that

$$w(x) = \sum_{n=1}^{\infty}\frac{2w_0}{n\pi}(-1)^{n+1}\sin\frac{n\pi}{L}x.$$

(b) If we assume $y(x) = \displaystyle\sum_{n=1}^{\infty}B_n\sin\frac{n\pi}{L}x$ then

$$y^{(4)} = \sum_{n=1}^{\infty}\frac{n^4\pi^4}{L^4}B_n\sin\frac{n\pi}{L}x$$

and so the differential equation $EIy^{(4)} = w(x)$ gives

$$B_n = \frac{2w_0(-1)^{n+1}L^4}{EIn^5\pi^5}.$$

Thus

$$y(x) = \frac{2w_0 L^4}{EI\pi^5}\sum_{n=1}^{\infty}\frac{(-1)^{n+1}}{n^5}\sin\frac{n\pi}{L}x.$$

Exercises 11.4

3. For $\lambda = 0$ the solution of $y'' = 0$ is $y = c_1 x + c_2$. The condition $y'(0) = 0$ implies $c_1 = 0$, so $\lambda = 0$ is an eigenvalue with corresponding eigenfunction 1.

For $\lambda < 0$ we have $y = c_1\cosh\sqrt{-\lambda}x + c_2\sin\sqrt{-\lambda}x$ and $y' = c_1\sqrt{-\lambda}\sinh\sqrt{-\lambda}x + c_2\sqrt{-\lambda}\cosh\sqrt{-\lambda}x$. The condition $y'(0) = 0$ implies $c_2 = 0$ and so $y = c_1\cosh\sqrt{-\lambda}x$. Now the condition $y'(L) = 0$ implies $c_1 = 0$. Thus $y = 0$ and there are no negative eigenvalues.

For $\lambda > 0$ we have $y = c_1\cos\sqrt{\lambda}x + c_2\sin\sqrt{\lambda}x$ and $y' = -c_1\sqrt{\lambda}\sin\sqrt{\lambda}x + c_2\sqrt{\lambda}\cos\sqrt{\lambda}x$. The condition $y'(0) = 0$ implies $c_2 = 0$ and so $y = c_1\cos\sqrt{\lambda}x$. Now the condition $y'(L) = 0$ implies $-c_1\sqrt{\lambda}\sin\sqrt{\lambda}L = 0$. For $c_1 \neq 0$ this condition will hold when $\sqrt{\lambda}L = n\pi$ or $\lambda = n^2\pi^2/L^2$, where $n = 1, 2, 3, \ldots$. These are the positive eigenvalues with corresponding eigenfunctions $\cos(n\pi/L)x$, $n = 1, 2, 3, \ldots$.

6. The eigenfunctions are $\sin\sqrt{\lambda_n}\,x$ where $\tan\sqrt{\lambda_n} = -\lambda_n$. Thus

$$\|\sin\sqrt{\lambda_n}\,x\|^2 = \int_0^1 \sin^2\sqrt{\lambda_n}\,x\,dx = \frac{1}{2}\int_0^1 \left(1 - \cos 2\sqrt{\lambda_n}\,x\right)dx$$

$$= \frac{1}{2}\left(x - \frac{1}{2\sqrt{\lambda_n}}\sin 2\sqrt{\lambda_n}\,x\right)\Big|_0^1 = \frac{1}{2}\left(1 - \frac{1}{2\sqrt{\lambda_n}}\sin 2\sqrt{\lambda_n}\right)$$

$$= \frac{1}{2}\left[1 - \frac{1}{2\sqrt{\lambda_n}}\left(2\sin\sqrt{\lambda_n}\cos\sqrt{\lambda_n}\right)\right]$$

$$= \frac{1}{2}\left[1 - \frac{1}{\sqrt{\lambda_n}}\tan\sqrt{\lambda_n}\cos\sqrt{\lambda_n}\cos\sqrt{\lambda_n}\right]$$

$$= \frac{1}{2}\left[1 - \frac{1}{\sqrt{\lambda_n}}\left(-\sqrt{\lambda_n}\cos^2\sqrt{\lambda_n}\right)\right] = \frac{1}{2}\left(1 + \cos^2\sqrt{\lambda_n}\right).$$

9. To obtain the self-adjoint form we note that an integrating factor is $(1/x)e^{\int(1-x)dx/x} = e^{-x}$. Thus, the differential equation is

$$xe^{-x}y'' + (1-x)e^{-x}y' + ne^{-x}y = 0$$

and the self-adjoint form is

$$\frac{d}{dx}\left[xe^{-x}y'\right] + ne^{-x}y = 0.$$

Identifying the weight function $p(x) = e^{-x}$ and noting that since $r(x) = xe^{-x}$, $r(0) = 0$ and $\lim_{x\to\infty} r(x) = 0$, we have the orthogonality relation

$$\int_0^\infty e^{-x}L_n(x)L_m(x)\,dx = 0, \ m \neq n.$$

12. (a) This is the parametric Bessel equation with $\nu = 1$. The general solution is

$$y = c_1 J_1(\lambda x)c_2 Y_1(\lambda x).$$

Since Y is bounded at 0 we must have $c_2 = 0$, so that $y = c_1 J_1(\lambda x)$. The condition $J_1(3\lambda) = 0$ defines the eigenvalues $\lambda_1, \lambda_2, \lambda_3, \ldots$. (When $\lambda = 0$ the differential equation is Cauchy-Euler and the only solution satisfying the boundary condition is $y = 0$, so $\lambda = 0$ is not an eigenvalue.)

(b) From Table 6.1 in Section 6.3 in the text we see that eigenvalues are determined by $3\lambda_1 = 3.832$, $3\lambda_2 = 7.016$, $3\lambda_3 = 10.173$, and $3\lambda_4 = 13.323$. The first four eigenvalues are thus $\lambda_1 = 1.2773$, $\lambda_2 = 2.3387$, $\lambda_3 = 3.391$, and $\lambda_4 = 4.441$.

Exercises 11.5

3. The boundary condition indicates that we use (15) and (16) in the text. With $b = 2$ we obtain

$$c_i = \frac{2}{4J_1^2(2\lambda_i)} \int_0^2 x J_0(\lambda_i x)\, dx$$

$$\boxed{t = \lambda_i x \qquad dt = \lambda_i\, dx}$$

$$= \frac{1}{2J_1^2(2\lambda_i)} \cdot \frac{1}{\lambda_i^2} \int_0^{2\lambda_i} t J_0(t)\, dt$$

$$= \frac{1}{2\lambda_i^2 J_1^2(2\lambda_i)} \int_0^{2\lambda_i} \frac{d}{dt}[t J_1(t)]\, dt \qquad \text{[From (4) in the text]}$$

$$= \frac{1}{2\lambda_i^2 J_1^2(2\lambda_i)} t J_1(t) \Big|_0^{2\lambda_i}$$

$$= \frac{1}{\lambda_i J_1(2\lambda_i)}.$$

Thus

$$f(x) = \sum_{i=1}^{\infty} \frac{1}{\lambda_i J_1(2\lambda_i)} J_0(\lambda_i x).$$

6. Writing the boundary condition in the form

$$2J_0(2\lambda) + 2\lambda J_0'(2\lambda) = 0$$

we identify $b = 2$ and $h = 2$. Using (17) and (18) in the text we obtain

$$c_i = \frac{2\lambda_i^2}{(4\lambda_i^2 + 4)J_0^2(2\lambda_i)} \int_0^2 x J_0(\lambda_i x)\, dx$$

$$\boxed{t = \lambda_i x \qquad dt = \lambda_i\, dx}$$

$$= \frac{\lambda_i^2}{2(\lambda_i^2 + 1)J_0^2(2\lambda_i)} \cdot \frac{1}{\lambda_i^2} \int_0^{2\lambda_i} t J_0(t)\, dt$$

$$= \frac{1}{2(\lambda_i^2 + 1)J_0^2(2\lambda_i)} \int_0^{2\lambda_i} \frac{d}{dt}[t J_1(t)]\, dt \qquad \text{[From (4) in the text]}$$

$$= \frac{1}{2(\lambda_i^2 + 1)J_0^2(2\lambda_i)} t J_1(t) \Big|_0^{2\lambda_i}$$

$$= \frac{\lambda_i J_1(2\lambda_i)}{(\lambda_i^2 + 1)J_0^2(2\lambda_i)}.$$

Thus

$$f(x) = \sum_{i=1}^{\infty} \frac{\lambda_i J_1(2\lambda_i)}{(\lambda_i^2 + 1)J_0^2(2\lambda_i)} J_0(\lambda_i x).$$

9. The boundary condition indicates that we use (19) and (20) in the text. With $b = 3$ we obtain

$$c_1 = \frac{2}{9} \int_0^3 x x^2 \, dx = \frac{2}{9} \frac{x^4}{4} \Big|_0^3 = \frac{9}{2},$$

$$c_i = \frac{2}{9 J_0^2(3\lambda_i)} \int_0^3 x J_0(\lambda_i x) x^2 \, dx$$

$$\boxed{t = \lambda_i x \qquad dt = \lambda_i \, dx}$$

$$= \frac{2}{9 J_0^2(3\lambda_i)} \cdot \frac{1}{\lambda_i^4} \int_0^{3\lambda_i} t^3 J_0(t) \, dt$$

$$= \frac{2}{9 \lambda_i^4 J_0^2(3\lambda_i)} \int_0^{3\lambda_i} t^2 \frac{d}{dt} [t J_1(t)] \, dt$$

$$\boxed{\begin{array}{ll} u = t^2 & dv = \frac{d}{dt}[t J_1(t)] \, dt \\ du = 2t \, dt & v = t J_1(t) \end{array}}$$

$$= \frac{2}{9 \lambda_i^4 J_0^2(3\lambda_i)} \left(t^3 J_1(t) \Big|_0^{3\lambda_i} - 2 \int_0^{3\lambda_i} t^2 J_1(t) \, dt \right)$$

With $n = 0$ in equation (5) in the text we have $J_0'(x) = -J_1(x)$, so the boundary condition $J_0'(3\lambda_i) = 0$ implies $J_1(3\lambda_i) = 0$. Then

$$c_i = \frac{2}{9 \lambda_i^4 J_0^2(3\lambda_i)} \left(-2 \int_0^{3\lambda_i} \frac{d}{dt} [t^2 J_2(t)] \, dt \right) = \frac{2}{9 \lambda_i^4 J_0^2(3\lambda_i)} \left(-2 t^2 J_2(t) \Big|_0^{3\lambda_i} \right)$$

$$= \frac{2}{9 \lambda_i^4 J_0^2(3\lambda_i)} \left[-18 \lambda_i^2 J_2(3\lambda_i) \right] = \frac{-4 J_2(3\lambda_i)}{\lambda_i^2 J_0^2(3\lambda_i)}.$$

Thus

$$f(x) = \frac{9}{2} - 4 \sum_{i=1}^{\infty} \frac{J_2(3\lambda_i)}{\lambda_i^2 J_0^2(3\lambda_i)} J_0(\lambda_i x).$$

15. Using $\cos^2 \theta = \frac{1}{2}(\cos 2\theta + 1)$ we have

$$P_2(\cos \theta) = \frac{1}{2}(3 \cos^2 \theta - 1) = \frac{3}{2} \cos^2 \theta - \frac{1}{2}$$

$$= \frac{3}{4}(\cos 2\theta + 1) - \frac{1}{2} = \frac{3}{4} \cos 2\theta + \frac{1}{4} = \frac{1}{4}(3 \cos 2\theta + 1).$$

18. If f is an odd function on $(-1, 1)$ then

$$\int_{-1}^1 f(x) P_{2n}(x) \, dx = 0$$

and

$$\int_{-1}^{1} f(x) P_{2n+1}(x)\, dx = 2 \int_{0}^{1} f(x) P_{2n+1}(x)\, dx.$$

Thus

$$c_{2n+1} = \frac{2(2n+1)+1}{2} \int_{-1}^{1} f(x) P_{2n+1}(x)\, dx = \frac{4n+3}{2} \left(2 \int_{0}^{1} f(x) P_{2n+1}(x)\, dx \right)$$

$$= (4n+1) \int_{0}^{1} f(x) P_{2n+1}(x)\, dx,$$

$c_{2n} = 0$, and

$$f(x) = \sum_{n=0}^{\infty} c_{2n+1} P_{2n+1}(x).$$

Chapter 11 Review Exercises

3. Cosine, since f is even

6. Periodically extending the function we see that at $x = -1$ the function converges to $\frac{1}{2}(-1+0) = -\frac{1}{2}$; at $x = 0$ it converges to $\frac{1}{2}(0+1) = \frac{1}{2}$, and at $x = 1$ it converges to $\frac{1}{2}(-1+0) = -\frac{1}{2}$.

9. Since the coefficient of y in the differential equation is n^2, the weight function is the integrating factor

$$\frac{1}{a(x)} e^{\int (b/a)dx} = \frac{1}{1-x^2} e^{\int -\frac{x}{1-x^2}\, dx} = \frac{1}{1-x^2} e^{\frac{1}{2}\ln(1-x^2)} = \frac{\sqrt{1-x^2}}{1-x^2} = \frac{1}{\sqrt{1-x^2}}$$

on the interval $[-1, 1]$. The orthogonality relation is

$$\int_{-1}^{1} \frac{1}{\sqrt{1-x^2}} T_m(x) T_n(x)\, dx = 0, \quad m \neq n.$$

12. (a) For $m \neq n$

$$\int_{0}^{L} \sin \frac{(2n+1)\pi}{2L} x \sin \frac{(2m+1)\pi}{2L} x\, dx = \frac{1}{2} \int_{0}^{L} \left(\cos \frac{n-m}{L} \pi x - \cos \frac{n+m+\pi}{L} \pi x \right) dx = 0.$$

(b) From

$$\int_{0}^{L} \sin^2 \frac{(2n+1)\pi}{2L} x\, dx = \int_{0}^{L} \left(\frac{1}{2} - \frac{1}{2} \cos \frac{(2n+1)\pi}{2L} x \right) dx = \frac{L}{2}$$

we see that

$$\left\| \sin \frac{(2n+1)\pi}{2L} x \right\| = \sqrt{\frac{L}{2}}.$$

15. (a) Since

$$A_0 = 2 \int_{0}^{1} e^{-x} dx$$

and

$$A_n = 2 \int_{-1}^{1} e^{-x} \cos n\pi x \, dx = \frac{2}{1 + n^2\pi^2}[(1 - (-1)^n e^{-1}]$$

for $n = 1, 2, 3, \ldots$ we have

$$f(x) = 1 - e^{-1} + 2 \sum_{n=1}^{\infty} \frac{1 - (-1)^n e^{-1}}{1 + n^2\pi^2} \cos n\pi x.$$

(b) Since

$$B_n = 2 \int_{0}^{1} e^{-x} \sin n\pi x \, dx = \frac{2n\pi}{1 + n^2\pi^2}[(1 - (-1)^n e^{-1}]$$

for $n = 1, 2, 3, \ldots$ we have

$$f(x) = \sum_{n=1}^{\infty} \frac{2n\pi}{1 + n^2\pi^2}[(1 - (-1)^n e^{-1}] \sin n\pi x.$$

18. To obtain the self-adjoint form of the differential equation in Problem 17 we note that an integrating factor is $(1/x^2)e^{\int dx/x} = 1/x$. Thus the weight function is $9/x$ and an orthogonality relation is

$$\int_{1}^{e} \frac{9}{x} \cos\left(\frac{2n-1}{2}\pi \ln x\right) \cos\left(\frac{2m-1}{2}\pi \ln x\right) dx = 0, \ m \neq n.$$

12 Partial Differential Equations and Boundary-Value Problems in Rectangular Coordinates

3. If $u = XY$ then

$$u_x = X'Y,$$
$$u_y = XY',$$
$$X'Y = X(Y - Y'),$$

and

$$\frac{X'}{X} = \frac{Y - Y'}{Y} = \pm\lambda^2.$$

Then

$$X' \mp \lambda^2 X = 0 \quad \text{and} \quad Y' - (1 \mp \lambda^2)Y = 0$$

so that

$$X = A_1 e^{\pm\lambda^2 x},$$
$$Y = A_2 e^{(1 \mp \lambda^2)y},$$

and

$$u = XY = c_1 e^{y + c_2(x - y)}.$$

6. If $u = XY$ then

$$u_x = X'Y,$$
$$u_y = XY',$$
$$yX'Y = xXY',$$

and

$$\frac{X'}{xX} = \frac{Y'}{-yY} = \pm\lambda^2.$$

Then

$$X \mp \lambda^2 x X = 0 \quad \text{and} \quad Y' \pm \lambda^2 y Y = 0$$

147

so that

$$X = A_1 e^{\pm \lambda^2 x^2 / 2},$$

$$Y = A_2 e^{\mp \lambda^2 y^2 / 2},$$

and

$$u = XY = c_1 e^{c_2 (x^2 - y^2)}.$$

9. If $u = XT$ then

$$u_t = XT',$$

$$u_{xx} = X''T,$$

$$kX''T - XT = XT',$$

and we choose

$$\frac{T'}{T} = \frac{kX'' - X}{X} = -1 \pm k\lambda^2$$

so that

$$T' - (-1 \pm k\lambda^2)T = 0 \quad \text{and} \quad X'' - (\pm \lambda^2)X = 0.$$

For $\lambda^2 > 0$ we obtain

$$X = A_1 \cosh \lambda x + A_2 \sinh \lambda x \quad \text{and} \quad T = A_3 e^{(-1 + k\lambda^2)t}$$

so that

$$u = XT = e^{(-1 + k\lambda^2)t} \left(c_1 \cosh \lambda x + c_2 \sinh \lambda x \right).$$

For $-\lambda^2 < 0$ we obtain

$$X = A_1 \cos \lambda x + A_2 \sin \lambda x \quad \text{and} \quad T = A_3 e^{(-1 - k\lambda^2)t}$$

so that

$$u = XT = e^{(-1 - k\lambda^2)t}(c_3 \cos \lambda x + c_4 \sin \lambda x).$$

If $\lambda^2 = 0$ then

$$X'' = 0 \quad \text{and} \quad T' + T = 0,$$

and we obtain

$$X = A_1 x + A_2 \quad \text{and} \quad T = A_3 e^{-t}.$$

In this case

$$u = XT = e^{-t}(c_5 x + c_6)$$

12. If $u = XT$ then

$$u_t = XT',$$

148

$$u_{tt} = XT'',$$

$$u_{xx} = X''T,$$

$$a^2 X''T = XT'' + 2kXT',$$

and

$$\frac{X''}{X} = \frac{T'' + 2kT'}{a^2 T} = \pm\lambda^2$$

so that

$$X'' \mp \lambda^2 X = 0 \quad \text{and} \quad T'' + 2kT' \mp a^2\lambda^2 T = 0.$$

For $\lambda^2 > 0$ we obtain

$$X = A_1 e^{\lambda x} + A_2 e^{-\lambda x},$$

$$T = A_3 e^{(-k+\sqrt{k^2+a^2\lambda^2}\,)t} + A_4 e^{(-k-\sqrt{k^2+a^2\lambda^2}\,)t},$$

and

$$u = XT = \left(A_1 e^{\lambda x} + A_2 e^{-\lambda x}\right)\left(A_3 e^{(-k+\sqrt{k^2+a^2\lambda^2}\,)t} + A_4 e^{(-k-\sqrt{k^2+a^2\lambda^2}\,)t}\right).$$

For $-\lambda^2 < 0$ we obtain

$$X = A_1 \cos \lambda x + A_2 \sin \lambda x.$$

If $k^2 - a^2\lambda^2 > 0$ then

$$T = A_3 e^{(-k+\sqrt{k^2-a^2\lambda^2}\,)t} + A_4 e^{(-k-\sqrt{k^2-a^2\lambda^2}\,)t}.$$

If $k^2 - a^2\lambda^2 < 0$ then

$$T = e^{-kt}\left(A_3 \cos\sqrt{a^2\lambda^2 - k^2}\,t + A_4 \sin\sqrt{a^2\lambda^2 - k^2}\,t\right).$$

If $k^2 - a^2\lambda^2 = 0$ then

$$T = A_3 e^{-kt} + A_4 t e^{-kt}$$

so that

$$u = XT = (A_1 \cos \lambda x + A_2 \sin \lambda x)\left(A_3 e^{(-k+\sqrt{k^2-a^2\lambda^2}\,)t} + A_4 e^{(-k-\sqrt{k^2-a^2\lambda^2}\,)t}\right)$$

$$= (A_1 \cos \lambda x + A_2 \sin \lambda x)e^{-kt}\left(A_3 \cos\sqrt{a^2\lambda^2 - k^2}\,t + A_4 \sin\sqrt{a^2\lambda^2 - k^2}\,t\right)$$

$$= \left(A_1 \cos\frac{k}{a}x + A_2 \sin\frac{k}{a}x\right)\left(A_3 e^{-kt} + A_4 t e^{-kt}\right).$$

For $\lambda^2 = 0$ we obtain

$$X = A_1 x + A_2,$$

$$T = A_3 + A_4 e^{-2kt},$$

and

$$u = XT = (A_1x + A_2)(A_3 + A_4e^{-2kt}).$$

15. If $u = XY$ then

$$u_{xx} = X''Y,$$

$$u_{yy} = XY'',$$

$$X''Y + XY'' = XY,$$

and

$$\frac{X''}{X} = \frac{Y - Y''}{Y} = \pm\lambda^2$$

so that

$$X'' \mp \lambda^2 X = 0 \quad \text{and} \quad Y'' + (\pm\lambda^2 - 1)Y = 0.$$

For $\lambda^2 > 0$ we obtain

$$X = A_1e^{\lambda x} + A_2e^{-\lambda x}.$$

If $\lambda^2 - 1 > 0$ then

$$Y = A_3 \cos \sqrt{\lambda^2 - 1}\, y + A_4 \sin \sqrt{\lambda^2 - 1}\, y.$$

If $\lambda^2 - 1 < 0$ then

$$Y = A_3e^{\sqrt{1-\lambda^2}\, y} + A_4e^{-\sqrt{1-\lambda^2}\, y}.$$

If $\lambda^2 - 1 = 0$ then $Y = A_3y + A_4$ so that

$$u = XY = \left(A_1e^{\lambda x} + A_2e^{-\lambda x}\right)\left(A_3 \cos \sqrt{\lambda^2 - 1}\, y + A_4 \sin \sqrt{\lambda^2 - 1}\, y\right),$$

$$= \left(A_1e^{\lambda x} + A_2e^{-\lambda x}\right)\left(A_3e^{\sqrt{1-\lambda^2}\, y} + A_4e^{-\sqrt{1-\lambda^2}\, y}\right)$$

$$= (A_1e^x + A_2e^{-x})(A_3y + A_4).$$

For $-\lambda^2 < 0$ we obtain

$$X = A_1 \cos \lambda x + A_2 \sin \lambda x,$$

$$Y = A_3e^{\sqrt{1+\lambda^2}\, y} + A_4e^{-\sqrt{1+\lambda^2}\, y},$$

and

$$u = XY = (A_1 \cos \lambda x + A_2 \sin \lambda x)\left(A_3e^{\sqrt{1+\lambda^2}\, y} + A_4e^{-\sqrt{1+\lambda^2}\, y}\right).$$

For $\lambda^2 = 0$ we obtain

$$X = A_1x + A_2,$$

$$Y = A_3e^y + A_4e^{-y},$$

and

$$u = XY = (A_1 x + A_2)(A_3 e^y + A_4 e^{-y}).$$

18. Identifying $A = 3$, $B = 5$, and $C = 1$, we compute $B^2 - 4AC = 13 > 0$. The equation is hyperbolic.

21. Identifying $A = 1$, $B = -9$, and $C = 0$, we compute $B^2 - 4AC = 81 > 0$. The equation is hyperbolic.

24. Identifying $A = 1$, $B = 0$, and $C = 1$, we compute $B^2 - 4AC = -4 < 0$. The equation is elliptic.

27. If $u = RT$ then

$$u_r = R'T,$$

$$u_{rr} = R''T,$$

$$u_t = RT',$$

$$RT' = k\left(R''T + \frac{1}{r}R'T\right),$$

and

$$\frac{r^2 R'' + rR'}{r^2 R} = \frac{T'}{kT} = \pm\lambda^2.$$

If we use $-\lambda^2 < 0$ then

$$r^2 R'' + rR' + \lambda^2 r^2 R = 0 \quad \text{and} \quad T'' \mp \lambda^2 kT = 0$$

so that

$$R = A_2 J_0(\lambda r) + A_3 Y_0(\lambda r),$$

$$T = A_1 e^{-k\lambda^2 t},$$

and

$$u = RT = e^{-k\lambda^2 t}[c_1 J_0(\lambda r) + c_2 Y_0(\lambda r)]$$

30. We identify $A = xy + 1$, $B = x + 2y$, and $C = 1$. Then $B^2 - 4AC = x^2 + 4y^2 - 4$. The equation $x^2 + 4y^2 = 4$ defines an ellipse. The partial differential equation is hyperbolic outside the ellipse, parabolic on the ellipse, and elliptic inside the ellipse.

Exercises 12.2

3. $k\dfrac{\partial^2 u}{\partial x^2} = \dfrac{\partial u}{\partial t}, \quad 0 < x < L, \ t > 0$

$u(0,t) = 100, \quad \left. \dfrac{\partial u}{\partial x} \right|_{x=L} = -hu(L,t), \quad t > 0$

$u(x,0) = f(x), \quad 0 < x < L$

6. $a^2\dfrac{\partial^2 u}{\partial x^2} = \dfrac{\partial^2 u}{\partial t^2}, \quad 0 < x < L, \ t > 0$

$u(0,t) = 0, \quad u(L,t) = 0, \quad t > 0$

$u(x,0) = 0, \quad \left. \dfrac{\partial u}{\partial x} \right|_{t=0} = \sin\dfrac{\pi x}{L}, \quad 0 < x < L$

9. $\dfrac{\partial^2 u}{\partial x^2} + \dfrac{\partial^2 u}{\partial y^2} = 0, \quad 0 < x < 4, \ 0 < y < 2$

$\left. \dfrac{\partial u}{\partial x} \right|_{x=0} = 0, \quad u(4,y) = f(y), \quad 0 < y < 2$

$\left. \dfrac{\partial u}{\partial y} \right|_{y=0} = 0, \quad u(x,2) = 0, \quad 0 < x < 4$

Exercises 12.3

3. Using $u = XT$ and $-\lambda^2$ as a separation constant leads to

$$X'' + \lambda^2 X = 0,$$

$$X'(0) = 0,$$

$$X'(L) = 0,$$

and

$$T' + k\lambda^2 T = 0.$$

Then

$$X = c_1 \cos\frac{n\pi}{L}x \quad \text{and} \quad T = c_2 e^{-\frac{kn^2\pi^2}{L^2}t}$$

for $n = 0, 1, 2, \ldots$ so that

$$u = \sum_{n=0}^{\infty} A_n e^{-\frac{kn^2\pi^2}{L^2}t} \cos\frac{n\pi}{L}x.$$

Imposing

$$u(x,0) = f(x) = \sum_{n=0}^{\infty} A_n \cos\frac{n\pi}{L}x$$

152

gives

$$u(x,t) = \frac{1}{L}\int_0^L f(x)\,dx + \frac{2}{L}\sum_{n=1}^{\infty}\left(\int_0^L f(x)\cos\frac{n\pi}{L}x\,dx\right)e^{-\frac{kn^2\pi^2}{L^2}t}\cos\frac{n\pi}{L}x.$$

6. Using $u = XT$ and $-\lambda^2$ as a separation constant leads to

$$X'' + \lambda^2 X = 0,$$

$$X(0) = 0,$$

$$X(L) = 0,$$

and

$$T' + (h + k\lambda^2)T = 0.$$

Then

$$X = c_1 \sin\frac{n\pi}{L}x \quad\text{and}\quad T = c_2 e^{-\left(h + \frac{kn^2\pi^2}{L^2}\right)t}$$

for $n = 1, 2, 3, \ldots$ so that

$$u = \sum_{n=1}^{\infty} A_n e^{-\left(h + \frac{kn^2\pi^2}{L^2}\right)t}\sin\frac{n\pi}{L}x.$$

Imposing

$$u(x,0) = f(x) = \sum_{n=1}^{\infty} A_n \sin\frac{n\pi}{L}x$$

gives

$$u = \frac{2}{L}\sum_{n=1}^{\infty}\left(\int_0^L f(x)\sin\frac{n\pi}{L}x\,dx\right)e^{-\left(h + \frac{kn^2\pi^2}{L^2}\right)t}\sin\frac{n\pi}{L}x.$$

Exercises 12.4

3. Using $u = XT$ and $-\lambda^2$ as a separation constant leads to

$$X'' + \lambda^2 X = 0,$$

$$X(0) = 0,$$

$$X(L) = 0,$$

and

$$T'' + \lambda^2 a^2 T = 0.$$

Then

$$X = c_1 \sin\frac{n\pi}{L}x \quad\text{and}\quad T = c_2 \cos\frac{n\pi a}{L}t + c_3 \sin\frac{n\pi a}{L}t.$$

for $n = 1, 2, 3, \ldots$ so that

$$u = \sum_{n=1}^{\infty} \left(A_n \cos \frac{n\pi a}{L}t + B_n \sin \frac{n\pi a}{L}t \right) \sin \frac{n\pi}{L}x.$$

Imposing

$$u(x, 0) = \sum_{n=1}^{\infty} A_n \sin \frac{n\pi}{L}x$$

gives

$$A_n = \frac{2}{L} \left(\int_0^{L/3} \frac{3}{L}x \sin \frac{n\pi}{L}x \, dx + \int_{L/3}^{2L/3} \sin \frac{n\pi}{L}x \, dx + \int_{2L/3}^{L} \left(3 - \frac{3}{L}x\right) \sin \frac{n\pi}{L}x \, dx \right)$$

so that

$$A_1 = \frac{6\sqrt{3}}{\pi^2},$$

$$A_2 = A_3 = A_4 = 0,$$

$$A_5 = -\frac{6\sqrt{3}}{5^2\pi^2},$$

$$A_6 = 0,$$

$$a_7 = \frac{6\sqrt{3}}{7^2\pi^2} \ldots.$$

Imposing

$$u_t(x, 0) = 0 = \sum_{n=1}^{\infty} B_n \frac{n\pi a}{L} \sin \frac{n\pi}{L}x$$

gives $B_n = 0$ for $n = 1, 2, 3, \ldots$ so that

$$u(x, t) = \frac{6\sqrt{3}}{\pi^2} \left(\cos \frac{\pi a}{L}t \sin \frac{\pi}{L}x - \frac{1}{5^2} \cos \frac{5\pi a}{L}t \sin \frac{5\pi}{L}x + \frac{1}{7^2} \cos \frac{7\pi a}{L}t \sin \frac{7\pi}{L}x - \cdots \right).$$

6. Using $u = XT$ and $-\lambda^2$ as a separation constant leads to

$$X'' + \lambda^2 X = 0,$$

$$X(0) = 0,$$

$$X(1) = 0,$$

and

$$T'' + \lambda^2 a^2 T = 0.$$

Then

$$X = c_1 \sin n\pi x \quad \text{and} \quad T = c_2 \cos n\pi at + c_3 \sin n\pi at$$

for $n = 1, 2, 3, \ldots$ so that

$$u = \sum_{n=1}^{\infty} (A_n \cos n\pi at + B_n \sin n\pi at) \sin n\pi x.$$

Imposing

$$u(x, 0) = 0.01 \sin 3\pi x = \sum_{n=1}^{\infty} A_n \sin n\pi x$$

and

$$u_t(x, 0) = 0 = \sum_{n=1}^{\infty} B_n n\pi a \sin n\pi x$$

gives $B_n = 0$ for $n = 1, 2, 3, \ldots$, $A_3 = 0.01$, and $A_n = 0$ for $n = 1, 2, 4, 5, 6, \ldots$ so that

$$u(x, t) = 0.01 \sin 3\pi x \cos 3\pi at.$$

9. Using $u = XT$ and $-\lambda^2$ as a separation constant leads to

$$X'' + \lambda^2 X = 0,$$

$$X(0) = 0,$$

$$X(\pi) = 0,$$

and

$$T'' + 2\beta T' + \lambda^2 T = 0.$$

Then

$$X = c_1 \sin nx \quad \text{and} \quad T = e^{-\beta t} \left(c_2 \cos \sqrt{n^2 - \beta^2}\, t + c_3 \sin \sqrt{n^2 - \beta^2}\, t \right)$$

so that

$$u = \sum_{n=1}^{\infty} e^{-\beta t} \left(A_n \cos \sqrt{n^2 - \beta^2}\, t + B_n \sin \sqrt{n^2 - \beta^2}\, t \right) \sin nx.$$

Imposing

$$u(x, 0) = f(x) = \sum_{n=1}^{\infty} A_n \sin nx$$

and

$$u_t(x, 0) = 0 = \sum_{n=1}^{\infty} \left(B_n \sqrt{n^2 - \beta^2} - \beta A_n \right) \sin nx$$

gives

$$u(x, t) = e^{-\beta t} \sum_{n=1}^{\infty} A_n \left(\cos \sqrt{n^2 - \beta^2}\, t + \frac{\beta}{\sqrt{n^2 - \beta^2}} \sin \sqrt{n^2 - \beta^2}\, t \right) \sin nx,$$

where

$$A_n = \frac{2}{\pi} \int_0^{\pi} f(x) \sin nx\, dx.$$

12. (a) Using

$$X = c_1 \cosh \lambda x + c_2 \sinh \lambda x + c_3 \cos \lambda x + c_4 \sin \lambda x$$

and $X(0) = 0$, $X'(0) = 0$ we find, in turn, $c_3 = -c_1$ and $c_4 = -c_2$. The conditions $X(L) = 0$ and $X'(L) = 0$ then yield the system of equations for c_1 and c_2:

$$c_1(\cosh \lambda L - \cos \lambda L) + c_2(\sinh \lambda L - \sin \lambda L) = 0$$

$$c_1(\lambda \sinh \lambda L + \lambda \sin \lambda L) + c_2(\lambda \cosh \lambda L - \lambda \cos \lambda L) = 0.$$

In order that this system have nontrivial solutions the determinant of the coefficients must be zero:

$$\lambda(\cosh \lambda L - \cos \lambda L)^2 - \lambda(\sinh^2 \lambda L - \sin^2 \lambda L) = 0.$$

$\lambda = 0$ is not an eigenvalue since this leads to $X = 0$. Thus the last equation simplifies to $\cosh \lambda L \cos \lambda L = 1$ or $\cosh x \cos x = 1$, where $x = \lambda L$.

(b) The equation $\cosh x \cos x = 1$ is the same as $\cos x = \operatorname{sech} x$. The figure indicates that the equation has an infinite number of roots.

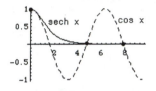

(c) Using a CAS we find the first four positive roots to be $x_1 = 4.7300$, $x_2 = 7.8532$, $x_3 = 10.9956$, and $x_4 = 14.1372$. Thus the first four eigenvalues are $\lambda_1 = x_1/L = 4.7300/L$, $\lambda_2 = x_2/L = 7.8532/L$, $\lambda_3 = x_3/L = 10.9956/L$, and $\lambda_4 = x_4/L = 14.1372/L$.

15. $u(x,t) = \dfrac{1}{2}[\sin(x + at) + \sin(x - at)] + \dfrac{1}{2a} \displaystyle\int_{x-at}^{x+at} ds$

$$= \frac{1}{2}[\sin x \cos at + \cos x \sin at + \sin x \cos at - \cos x \sin at] + \frac{1}{2a} s \Big|_{x-at}^{x+at} = \sin x \cos at + t$$

18.

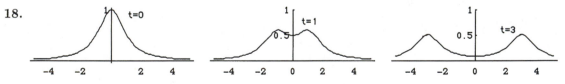

21. (a) and (b) With the given parameters, the solution is

$$u(x,t) = \frac{8}{\pi^2} \sum_{n=1}^{\infty} \frac{1}{n^2} \sin \frac{n\pi}{2} \cos nt \sin nx.$$

For n even, $\sin(n\pi/2) = 0$, so the first six nonzero terms correspond to $n = 1, 3, 5, 7, 9, 11$. In this case $\sin(n\pi/2) = \sin(2p - 1)/2 = (-1)^{p+1}$ for $p = 1, 2, 3, 4, 5, 6$, and

$$u(x,t) = \frac{8}{\pi^2} \sum_{p=1}^{\infty} \frac{(-1)^{p+1}}{(2p-1)^2} \cos(2p-1)t \sin(2p-1)x.$$

Frames of the movie corresponding to $t = 0.5, 1, 1.5$, and 2 are shown.

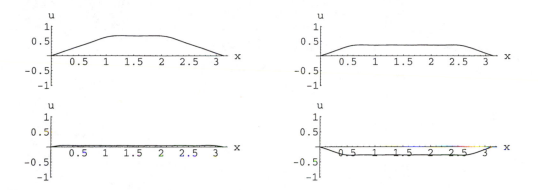

Exercises 12.5

3. Using $u = XY$ and $-\lambda^2$ as a separation constant leads to

$$X'' + \lambda^2 X = 0,$$

$$X(0) = 0,$$

$$X(a) = 0,$$

and

$$Y'' - \lambda^2 Y = 0,$$

$$Y(b) = 0.$$

Then

$$X = c_1 \sin \frac{n\pi}{a} x \quad \text{and} \quad Y = c_2 \cosh \frac{n\pi}{a} y - c_2 \frac{\cosh \frac{n\pi b}{a}}{\sinh \frac{n\pi b}{a}} \sinh \frac{n\pi}{a} y$$

for $n = 1, 2, 3, \ldots$ so that

$$u = \sum_{n=1}^{\infty} A_n \sin \frac{n\pi}{a} x \left(\cosh \frac{n\pi}{a} y - \frac{\cosh \frac{n\pi b}{a}}{\sinh \frac{n\pi b}{a}} \sinh \frac{n\pi}{a} y \right).$$

Imposing

$$u(x, 0) = f(x) = \sum_{n=1}^{\infty} A_n \sin \frac{n\pi}{a} x$$

gives

$$A_n = \frac{2}{a} \int_0^a f(x) \sin \frac{n\pi}{a} x \, dx$$

157

so that

$$u(x,y) = \frac{2}{a} \sum_{n=1}^{\infty} \left(\int_0^a f(x) \sin \frac{n\pi}{a} x \, dx \right) \sin \frac{n\pi}{a} x \left(\cosh \frac{n\pi}{a} y - \frac{\cosh \frac{n\pi b}{a}}{\sinh \frac{n\pi b}{a}} \sinh \frac{n\pi}{a} y \right).$$

6. Using $u = XY$ and λ^2 as a separation constant leads to

$$X'' - \lambda^2 X = 0,$$

$$X'(1) = 0,$$

and

$$Y'' + \lambda^2 Y = 0,$$

$$Y'(0) = 0,$$

$$Y'(\pi) = 0.$$

Then

$$Y = c_1 \cos ny$$

for $n = 0, 1, 2, \ldots$ and

$$X = c_2 \cosh nx - c_2 \frac{\sinh n}{\cosh n} \sinh nx$$

for $n = 0, 1, 2, \ldots$ so that

$$u = A_0 + \sum_{n=1}^{\infty} A_n \left(\cosh nx - \frac{\sinh n}{\cosh n} \sinh nx \right) \cos ny.$$

Imposing

$$u(0,y) = g(y) = A_0 + \sum_{n=1}^{\infty} A_n \cos ny$$

gives

$$A_0 = \frac{1}{\pi} \int_0^{\pi} g(y) \, dy \quad \text{and} \quad A_n = \frac{2}{\pi} \int_0^{\pi} g(y) \cos ny \, dy$$

for $n = 1, 2, 3, \ldots$ so that

$$u(x,y) = \frac{1}{\pi} \int_0^{\pi} g(y) \, dy + \sum_{n=1}^{\infty} \left(\frac{2}{\pi} \int_0^{\pi} g(y) \cos ny \, dy \right) \left(\cosh nx - \frac{\sinh n}{\cosh n} \sinh nx \right) \cos ny.$$

9. This boundary-value problem has the form of Problem 1 in this section in the text with $a = b = 1$, $f(x) = 100$, and $g(x) = 200$. The solution, then, is

$$u(x,y) = \sum_{n=1}^{\infty} (A_n \cosh n\pi y + B_n \sinh n\pi y) \sin n\pi x,$$

where

$$A_n = 2 \int_0^1 100 \sin n\pi x \, dx = 200 \left(\frac{1 - (-1)^n}{n\pi} \right)$$

and

$$B_n = \frac{1}{\sinh n\pi}\left[2\int_0^1 200\sin n\pi x\, dx - A_n\cosh n\pi\right]$$

$$= \frac{1}{\sinh n\pi}\left[400\left(\frac{1-(-1)^n}{n\pi}\right) - 200\left(\frac{1-(-1)^n}{n\pi}\right)\cosh n\pi\right]$$

$$= 200\left[\frac{1-(-1)^n}{n\pi}\right][2\operatorname{csch} n\pi - \coth n\pi].$$

12. Using $u = XY$ and $-\lambda^2$ as a separation constant leads to

$$X'' + \lambda^2 X = 0,$$

$$X'(0) = 0,$$

$$X'(\pi) = 0,$$

and

$$Y'' - \lambda^2 Y = 0.$$

By the boundedness of u as $y \to \infty$ we obtain $Y = c_1 e^{-ny}$ for $n = 1, 2, 3, \ldots$ or $Y = c_1$ and $X = c_2 \cos nx$ for $n = 0, 1, 2, \ldots$ so that

$$u = A_0 + \sum_{n=1}^{\infty} A_n e^{-ny}\cos nx.$$

Imposing

$$u(x,0) = f(x) = A_0 + \sum_{n=1}^{\infty} A_n \cos nx$$

gives

$$A_0 = \frac{1}{\pi}\int_0^\pi f(x)\, dx \quad \text{and} \quad A_n = \frac{2}{\pi}\int_0^\pi f(x)\cos nx\, dx$$

so that

$$u(x,y) = \frac{1}{\pi}\int_0^\pi f(x)\, dx + \sum_{n=1}^{\infty}\left(\frac{2}{\pi}\int_0^\pi f(x)\cos nx\, dx\right)e^{-ny}\cos nx.$$

15. In this problem we refer to the discussion in the text under the heading **Superposition Principle**. We identify $a = b = \pi$, $f(x) = 0$, $g(x) = 1$, $F(y) = 1$, and $G(y) = 1$. Then $A_n = 0$ and

$$u_1(x,y) = \sum_{n=1}^{\infty} B_n \sinh ny \sin nx$$

where

$$B_n = \frac{2}{\pi \sinh n\pi}\int_0^\pi \sin nx\, dx = \frac{2[1-(-1)^n]}{n\pi \sinh n\pi}.$$

Next

$$u_2(x,y) = \sum_{n=1}^{\infty}(A_n \cosh nx + B_n \sinh nx)\sin ny$$

where

$$A_n = \frac{2}{\pi} \int_0^\pi \sin ny \, dy = \frac{2[1 - (-1)^n]}{n\pi}$$

and

$$B_n = \frac{1}{\sinh n\pi} \left(\frac{2}{\pi} \int_0^\pi \sin ny \, dy - A_n \cosh n\pi \right)$$

$$= \frac{1}{\sinh n\pi} \left(\frac{2[1 - (-1)^n]}{n\pi} - \frac{2[1 - (-1)^n]}{n\pi} \cosh n\pi \right)$$

$$= \frac{2[1 - (-1)^n]}{n\pi \sinh n\pi} (1 - \cosh n\pi).$$

Now

$$A_n \cosh nx + B_n \sinh nx = \frac{2[1 - (-1)^n]}{n\pi} \left[\cosh nx + \frac{\sinh nx}{\sinh n\pi} (1 - \cosh n\pi) \right]$$

$$= \frac{2[1 - (-1)^n]}{n\pi \sinh n\pi} [\cosh nx \sinh n\pi + \sinh nx - \sinh nx \cosh n\pi]$$

$$= \frac{2[1 - (-1)^n]}{n\pi \sinh n\pi} [\sinh nx + \sinh n(\pi - x)]$$

and

$$u(x, y) = u_1 + u_2 = \frac{2}{\pi} \sum_{n=1}^\infty \frac{1 - (-1)^n}{n \sinh n\pi} \sinh ny \sin nx$$

$$+ \frac{2}{\pi} \sum_{n=1}^\infty \frac{[1 - (-1)^n][\sinh nx + \sinh n(\pi - x)]}{n \sinh n\pi} \sin ny.$$

18.

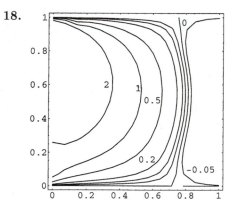

Exercises 12.6

3. If we let $u(x, t) = v(x, t) + \psi(x)$, then we obtain as in Example 1 in the text

$$k\psi'' + r = 0$$

or

$$\psi(x) = -\frac{r}{2k}x^2 + c_1 x + c_2.$$

The boundary conditions become

$$u(0, t) = v(0, t) + \psi(0) = u_0$$

$$u(1, t) = v(1, t) + \psi(1) = u_0.$$

Letting $\psi(0) = \psi(1) = u_0$ we obtain homogeneous boundary conditions in v:

$$v(0, t) = 0 \quad \text{and} \quad v(1, t) = 0.$$

Now $\psi(0) = \psi(1) = u_0$ implies $c_2 = u_0$ and $c_1 = r/2k$. Thus

$$\psi(x) = -\frac{r}{2k}x^2 + \frac{r}{2k}x + u_0 = u_0 - \frac{r}{2k}x(x - 1).$$

To determine $v(x, t)$ we solve

$$k\frac{\partial^2 v}{\partial x^2} = \frac{\partial v}{dt}, \quad 0 < x < 1, \ t > 0$$

$$v(0, t) = 0, \quad v(1, t) = 0,$$

$$v(x, 0) = \frac{r}{2k}x(x - 1) - u_0.$$

Separating variables, we find

$$v(x, t) = \sum_{n=1}^{\infty} A_n e^{-kn^2\pi^2 t} \sin n\pi x,$$

where

$$A_n = 2\int_0^1 \left[\frac{r}{2k}x(x - 1) - u_0\right]\sin n\pi x\, dx = 2\left[\frac{u_0}{n\pi} + \frac{r}{kn^3\pi^3}\right][(-1)^n - 1]. \tag{1}$$

Hence, a solution of the original problem is

$$u(x, t) = \psi(x) + v(x, t)$$

$$= u_0 - \frac{r}{2k}x(x - 1) + \sum_{n=1}^{\infty} A_n e^{-kn^2\pi^2 t} \sin n\pi x,$$

where A_n is defined in (1).

6. Substituting $u(x, t) = v(x, t) + \psi(x)$ into the partial differential equation gives

$$k\frac{\partial^2 v}{\partial x^2} + k\psi'' - hv - h\psi = \frac{\partial v}{\partial t}.$$

161

This equation will be homogeneous provided ψ satisfies

$$k\psi'' - h\psi = 0.$$

Since k and h are positive, the general solution of this latter equation is

$$\psi(x) = c_1 \cosh\sqrt{\frac{h}{k}}\,x + c_2 \sinh\sqrt{\frac{h}{k}}\,x.$$

From $\psi(0) = 0$ and $\psi(\pi) = u_0$ we find $c_1 = 0$ and $c_2 = u_0/\sinh\sqrt{h/k}\,\pi$. Hence

$$\psi(x) = u_0 \frac{\sinh\sqrt{h/k}\,x}{\sinh\sqrt{h/k}\,\pi}.$$

Now the new problem is

$$k\frac{\partial^2 v}{\partial x^2} - hv = \frac{\partial v}{\partial t}, \quad 0 < x < \pi,\ t > 0$$

$$v(0, t) = 0, \quad v(\pi, t) = 0, \quad t > 0$$

$$v(x, 0) = -\psi(x), \quad 0 < x < \pi.$$

If we let $v = XT$ then

$$\frac{X''}{X} = \frac{T' + hT}{kT} = -\lambda^2$$

gives the separated differential equations

$$X'' + \lambda^2 X = 0 \quad \text{and} \quad T' + \left(h + k\lambda^2\right)T = 0.$$

The respective solutions are

$$X(x) = c_3 \cos\lambda x + c_4 \sin\lambda x$$

$$T(t) = c_5 e^{-\left(h+k\lambda^2\right)t}.$$

From $X(0) = 0$ we get $c_3 = 0$ and from $X(\pi) = 0$ we find $\lambda = n$ for $n = 1, 2, 3, \ldots$. Consequently, it follows that

$$v(x, t) = \sum_{n=1}^{\infty} A_n e^{-\left(h+kn^2\right)t} \sin nx$$

where

$$A_n = -\frac{2}{\pi}\int_0^\pi \psi(x) \sin nx\, dx.$$

Hence a solution of the original problem is

$$u(x, t) = u_0 \frac{\sinh\sqrt{h/k}\,x}{\sinh\sqrt{h/k}\,\pi} + e^{-ht}\sum_{n=1}^{\infty} A_n e^{-kn^2 t} \sin nx$$

where

$$A_n = -\frac{2}{\pi} \int_0^\pi u_0 \frac{\sinh\sqrt{h/k}\,x}{\sinh\sqrt{h/k}\,\pi} \sin nx\, dx.$$

Using the exponential definition of the hyperbolic sine and integration by parts we find

$$A_n = \frac{2u_0 nk(-1)^n}{\pi\,(h+kn^2)}.$$

9. Substituting $u(x,t) = v(x,t) + \psi(x)$ into the partial differential equation gives

$$a^2 \frac{\partial^2 v}{\partial x^2} + a^2\psi'' + Ax = \frac{\partial^2 v}{\partial t^2}.$$

This equation will be homogeneous provided ψ satisfies

$$a^2\psi'' + Ax = 0.$$

The general solution of this differential equation is

$$\psi(x) = -\frac{A}{6a^2}x^3 + c_1 x + c_2.$$

From $\psi(0) = 0$ we obtain $c_2 = 0$, and from $\psi(1) = 0$ we obtain $c_1 = A/6a^2$. Hence

$$\psi(x) = \frac{A}{6a^2}(x - x^3).$$

Now the new problem is

$$a^2 \frac{\partial^2 v}{\partial x^2} = \frac{\partial^2 v}{\partial t^2}$$

$$v(0,t) = 0, \quad v(1,t) = 0, \quad t > 0,$$

$$v(x,0) = -\psi(x), \quad v_t(x,0) = 0, \quad 0 < x < 1.$$

Identifying this as the wave equation solved in Section 12.4 in the text with $L = 1$, $f(x) = -\psi(x)$, and $g(x) = 0$ we obtain

$$v(x,t) = \sum_{n=1}^\infty A_n \cos n\pi at \sin n\pi x$$

where

$$A_n = 2\int_0^1 [-\psi(x)] \sin n\pi x\, dx = \frac{A}{3a^2}\int_0^1 (x^3 - x)\sin n\pi x\, dx = \frac{2A(-1)^n}{a^2\pi^3 n^3}.$$

Thus

$$u(x,t) = \frac{A}{6a^2}(x - x^3) + \frac{2A}{a^2\pi^3}\sum_{n=1}^\infty \frac{(-1)^n}{n^3}\cos n\pi at \sin n\pi x.$$

12. Substituting $u(x,y) = v(x,y) + \psi(x)$ into Poisson's equation we obtain

$$\frac{\partial^2 v}{\partial x^2} + \psi''(x) + h + \frac{\partial^2 v}{\partial y^2} = 0.$$

The equation will be homogeneous provided ψ satisfies $\psi''(x) + h = 0$ or $\psi(x) = -\frac{h}{2}x^2 + c_1 x + c_2$. From $\psi(0) = 0$ we obtain $c_2 = 0$. From $\psi(\pi) = 1$ we obtain

$$c_1 = \frac{1}{\pi} + \frac{h\pi}{2}.$$

Then

$$\psi(x) = \left(\frac{1}{\pi} + \frac{h\pi}{2}\right)x - \frac{h}{2}x^2.$$

The new boundary-value problem is

$$\frac{\partial^2 v}{\partial x^2} + \frac{\partial^2 v}{\partial y^2} = 0$$

$$v(0, y) = 0, \quad v(\pi, y) = 0,$$

$$v(x, 0) = -\psi(x), \quad 0 < x < \pi.$$

Using $v(x, y) = XY$ and $-\lambda^2$ as a separation constant leads to

$$X'' + \lambda^2 X = 0,$$

$$X(0) = 0,$$

$$X(\pi) = 0,$$

and

$$Y'' - \lambda^2 Y = 0.$$

Then the boundedness of v as $y \to \infty$ gives $Y = c_1 e^{-ny}$ and $X = c_2 \sin nx$ for $n = 1, 2, 3, \ldots$ so that

$$v(x, y) = \sum_{n=1}^{\infty} A_n e^{-ny} \sin nx$$

where

$$A_n = \frac{2}{\pi} \int_0^{\pi} [-\psi(x) \sin nx] \, dx$$

$$= \frac{2(-1)^n}{m}\left(\frac{1}{\pi} + \frac{h\pi}{2}\right) - h(-1)^n \left(\frac{\pi}{n} + \frac{2}{n^2}\right).$$

Thus

$$u(x, y) = v(x, y) + \psi(x) = \left(\frac{1}{\pi} + \frac{h\pi}{2}\right)x - \frac{h}{2}x^2 + \sum_{n=1}^{\infty} A_n e^{-ny} \sin nx.$$

Exercises 12.7

3. Separating variables in Laplace's equation gives

$$X'' + \lambda^2 X = 0$$

$$Y'' - \lambda^2 Y = 0$$

and

$$X(x) = c_1 \cos \lambda x + c_2 \sin \lambda x$$

$$Y(y) = c_3 \cosh \lambda y + c_4 \sinh \lambda y.$$

From $u(0, y) = 0$ we obtain $X(0) = 0$ and $c_1 = 0$. From $u_x(a, y) = -hu(a, y)$ we obtain $X'(a) = -hX(a)$ and

$$\lambda \cos \lambda a = -h \sin \lambda a \quad \text{or} \quad \tan \lambda a = -\frac{\lambda}{h}.$$

Let λ_n, where $n = 1, 2, 3, \ldots$, be the consecutive positive roots of this equation. From $u(x, 0) = 0$ we obtain $Y(0) = 0$ and $c_3 = 0$. Thus

$$u(x, y) = \sum_{n=1}^{\infty} A_n \sinh \lambda_n y \sin \lambda_n x.$$

Now

$$f(x) = \sum_{n=1}^{\infty} A_n \sinh \lambda_n b \sin \lambda_n x$$

and

$$A_n \sinh \lambda_n b = \frac{\int_0^a f(x) \sin \lambda_n x \, dx}{\int_0^a \sin^2 \lambda_n x \, dx}.$$

Since

$$\int_0^a \sin^2 \lambda_n x \, dx = \frac{1}{2} \left[a - \frac{1}{2\lambda_n} \sin 2\lambda_n a \right] = \frac{1}{2} \left[a - \frac{1}{\lambda_n} \sin \lambda_n a \cos \lambda_n a \right]$$

$$= \frac{1}{2} \left[a - \frac{1}{h\lambda_n} (h \sin \lambda_n a) \cos \lambda_n a \right]$$

$$= \frac{1}{2} \left[a - \frac{1}{h\lambda_n} (-\lambda_n \cos \lambda_n a) \cos \lambda_n a \right] = \frac{1}{2h} \left[ah + \cos^2 \lambda_n a \right],$$

we have

$$A_n = \frac{2h}{\sinh \lambda_n b [ah + \cos^2 \lambda_n a]} \int_0^a f(x) \sin \lambda_n x \, dx.$$

6. Substituting $u(x, t) = v(x, t) + \psi(x)$ into the partial differential equation gives

$$a^2 \frac{\partial^2 v}{\partial x^2} + \psi''(x) = \frac{\partial^2 v}{\partial t^2}.$$

This equation will be homogeneous if $\psi''(x) = 0$ or $\psi(x) = c_1 x + c_2$. The boundary condition $u(0, t) = 0$ implies $\psi(0) = 0$ which implies $c_2 = 0$. Thus $\psi(x) = c_1 x$. Using the second boundary condition, we obtain

$$E\left(\frac{\partial v}{\partial x} + \psi'\right)\bigg|_{x=L} = F_0,$$

which will be homogeneous when

$$E\psi'(L) = F_0.$$

Since $\psi'(x) = c_1$ we conclude that $c_1 = F_0/E$ and

$$\psi(x) = \frac{F_0}{E}x.$$

The new boundary-value problem is

$$a^2 \frac{\partial^2 v}{\partial x^2} = \frac{\partial^2 v}{\partial t^2}, \quad 0 < x < L, \quad t > 0$$

$$v(0, t) = 0, \quad \frac{\partial v}{\partial x}\bigg|_{x=L} = 0, \quad t > 0,$$

$$v(x, 0) = -\frac{F_0}{E}x, \quad \frac{\partial v}{\partial t}\bigg|_{t=0} = 0, \quad 0 < x < L.$$

Referring to Example 2 in the text we see that

$$v(x, t) = \sum_{n=1}^{\infty} A_n \cos a\left(\frac{2n-1}{2L}\right)\pi t \sin\left(\frac{2n-1}{2L}\right)\pi x$$

where

$$-\frac{F_0}{E}x = \sum_{n=1}^{\infty} A_n \sin\left(\frac{2n-1}{2L}\right)\pi x$$

and

$$A_n = \frac{-F_0 \int_0^L x \sin\left(\frac{2n-1}{2L}\right)\pi x \, dx}{E \int_0^L \sin^2\left(\frac{2n-1}{2L}\right)\pi x \, dx} = \frac{8F_0 L(-1)^n}{E\pi^2(2n-1)^2}.$$

Thus

$$u(x, t) = v(x, t) + \psi(x)$$

$$= \frac{F_0}{E}x + \frac{8F_0 L}{E\pi^2} \sum_{n=1}^{\infty} \frac{(-1)^n}{(2n-1)^2} \cos a\left(\frac{2n-1}{2L}\right)\pi t \sin\left(\frac{2n-1}{2L}\right)\pi x.$$

9. (a) Using $u = XT$ and separation constant λ^4 we find

$$X^{(4)} - \lambda^4 X = 0$$

and

$$X(x) = c_1 \cos \lambda x + c_2 \sin \lambda x + c_3 \cosh \lambda x + c_4 \sinh \lambda x..$$

Since $u = XT$ the boundary conditions become

$$X(0) = 0, \quad X'(0) = 0, \quad X''(1) = 0, \quad X'''(1) = 0.$$

Now $X(0) = 0$ implies $c_1 + c_3 = 0$, while $X'(0) = 0$ implies $c_2 + c_4 = 0$. Thus

$$X(x) = c_1 \cos \lambda x + c_2 \sin \lambda x - c_1 \cosh \lambda x - c_2 \sinh \lambda x.$$

The boundary condition $X''(1) = 0$ implies

$$-c_1 \cos \lambda - c_2 \sin \lambda - c_1 \cosh \lambda - c_2 \sinh \lambda = 0$$

while the boundary condition $X'''(1) = 0$ implies

$$c_1 \sin \lambda - c_2 \cos \lambda - c_1 \sinh \lambda - c_2 \cosh \lambda = 0.$$

We then have the system of two equations in two unknowns

$$(\cos \lambda + \cosh \lambda)c_1 + (\sin \lambda + \sinh \lambda)c_2 = 0$$

$$(\sin \lambda - \sinh \lambda)c_1 - (\cos \lambda + \cosh \lambda)c_2 = 0.$$

This homogeneous system will have nontrivial solutions for c_1 and c_2 provided

$$\begin{vmatrix} \cos \lambda + \cosh \lambda & \sin \lambda + \sinh \lambda \\ \sin \lambda - \sinh \lambda & -\cos \lambda - \cosh \lambda \end{vmatrix} = 0$$

or

$$-2 - 2 \cos \lambda \cosh \lambda = 0.$$

Thus, the eigenvalues are determined by the equation $\cos \lambda \cosh \lambda = -1$.

(b) Using a computer to graph $\cosh \lambda$ and $-1/\cos \lambda = -\sec \lambda$ we see that the first two positive eigenvalues occur near 1.9 and 4.7. Applying Newton's method with these initial values we find that the eigenvalues are $\lambda_1 = 1.8751$ and $\lambda_2 = 4.6941$.

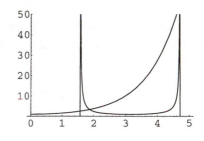

Exercises 12.8

3. In this problem we need to solve the partial differential equation

$$a^2 \left(\frac{\partial^2 u}{\partial x^2} + \frac{\partial^2 u}{\partial y^2} \right) = \frac{\partial^2 u}{\partial t^2}.$$

To separate this equation we try $u(x, y, t) = X(x)Y(y)T(t)$:

$$a^2 (X''YT + XY''T) = XYT''$$

$$\frac{X''}{X} = -\frac{Y''}{Y} + \frac{T''}{a^2T} = -\lambda^2.$$

Then

$$X'' + \lambda^2 X = 0 \tag{1}$$

$$\frac{Y''}{Y} = \frac{T''}{a^2T} + \lambda^2 = -\mu^2$$

$$Y'' + \mu^2 Y = 0 \tag{2}$$

$$T'' + a^2 \left(\lambda^2 + \mu^2 \right) T = 0. \tag{3}$$

The general solutions of equations (1), (2), and (3) are, respectively,

$$X(x) = c_1 \cos \lambda x + c_2 \sin \lambda x$$

$$Y(y) = c_3 \cos \mu y + c_4 \sin \mu y$$

$$T(t) = c_5 \cos a\sqrt{\lambda^2 + \mu^2}\, t + c_6 \sin a\sqrt{\lambda^2 + \mu^2}\, t.$$

The conditions $X(0) = 0$ and $Y(0) = 0$ give $c_1 = 0$ and $c_3 = 0$. The conditions $X(\pi) = 0$ and $Y(\pi) = 0$ yield two sets of eigenvalues:

$$\lambda = m, \ m = 1, 2, 3, \ldots \quad \text{and} \quad \mu = n, \ n = 1, 2, 3, \ldots .$$

A product solution of the partial differential equation that satisfies the boundary conditions is

$$u_{mn}(x, y, t) = \left(A_{mn} \cos a\sqrt{m^2 + n^2}\, t + B_{mn} \sin a\sqrt{m^2 + n^2}\, t \right) \sin mx \sin ny.$$

To satisfy the initial conditions we use the superposition principle:

$$u(x, y, t) = \sum_{m=1}^{\infty} \sum_{n=1}^{\infty} \left(A_{mn} \cos a\sqrt{m^2 + n^2}\, t + B_{mn} \sin a\sqrt{m^2 + n^2}\, t \right) \sin mx \sin ny.$$

The initial condition $u_t(x, y, 0) = 0$ implies $B_{mn} = 0$ and

$$u(x, y, t) = \sum_{m=1}^{\infty} \sum_{n=1}^{\infty} A_{mn} \cos a\sqrt{m^2 + n^2}\, t \sin mx \sin ny.$$

168

At $t = 0$ we have

$$xy(x - \pi)(y - \pi) = \sum_{m=1}^{\infty} \sum_{n=1}^{\infty} A_{mn} \sin mx \sin ny.$$

Using (11) and (12) in the text, it follows that

$$A_{mn} = \frac{4}{\pi^2} \int_0^\pi \int_0^\pi xy(x - \pi)(y - \pi) \sin mx \sin ny \, dx \, dy$$

$$= \frac{4}{\pi^2} \int_0^\pi x(x - \pi) \sin mx \, dx \int_0^\pi y(y - \pi) \sin ny \, dy$$

$$= \frac{16}{m^3 n^3 \pi^2} [(-1)^m - 1][(-1)^n - 1].$$

6. To separate Laplace's equation in three dimensions we try $u(x, y, z) = X(x)Y(y)Z(z)$:

$$X''YZ + XY''Z + XYZ'' = 0$$

$$\frac{X''}{X} = -\frac{Y''}{Y} - \frac{Z''}{Z} = -\lambda^2.$$

Then

$$X'' + \lambda^2 X = 0 \tag{4}$$

$$\frac{Y''}{Y} = -\frac{Z''}{Z} + \lambda^2 = -\mu^2$$

$$Y'' + \mu^2 Y = 0 \tag{5}$$

$$Z'' - (\lambda^2 + \mu^2)Z = 0. \tag{6}$$

The general solutions of equations (4), (5), and (6) are, respectively

$$X(x) = c_1 \cos \lambda x + c_2 \sin \lambda x$$

$$Y(y) = c_3 \cos \mu y + c_4 \sin \mu y$$

$$Z(z) = c_5 \cosh \sqrt{\lambda^2 + \mu^2}\, z + c_6 \sinh \sqrt{\lambda^2 + \mu^2}\, z.$$

The boundary and initial conditions are

$$u(0, y, z) = 0, \qquad u(a, y, z) = 0,$$

$$u(x, 0, z) = 0, \qquad u(x, b, z) = 0,$$

$$u(x, y, 0) = f(x, y), \qquad u(x, y, c) = 0.$$

The conditions $X(0) = Y(0) = 0$ give $c_1 = c_3 = 0$. The conditions $X(a) = Y(b) = 0$ yield two sets of eigenvalues:

$$\lambda = \frac{m\pi}{a}, \quad m = 1, 2, 3, \dots \quad \text{and} \quad \mu = \frac{n\pi}{b}, \quad n = 1, 2, 3, \dots .$$

169

Let

$$\omega_{mn}^2 = \frac{m^2\pi^2}{a^2} + \frac{n^2\pi^2}{b^2}.$$

Then the boundary condition $Z(c) = 0$ gives

$$c_5 \cosh c\omega_{mn} + c_6 \sinh c\omega_{mn} = 0$$

from which we obtain

$$Z(z) = c_5\left(\cosh \omega_{mn}z - \frac{\cosh c\omega_{mn}}{\sinh c\omega_{mn}}\sinh \omega z\right)$$

$$= \frac{c_5}{\sinh c\omega_{mn}}(\sinh c\omega_{mn}\cosh \omega_{mn}z - \cosh c\omega_{mn}\sinh \omega_{mn}z) = c_{mn}\sinh \omega_{mn}(c-z).$$

By the superposition principle

$$u(x,y,t) = \sum_{m=1}^{\infty}\sum_{n=1}^{\infty} A_{mn}\sinh \omega_{mn}(c-z)\sin\frac{m\pi}{a}x\sin\frac{n\pi}{b}y$$

where

$$A_{mn} = \frac{4}{ab\sinh c\omega_{mn}}\int_0^b\int_0^a f(x,y)\sin\frac{m\pi}{a}x\sin\frac{n\pi}{b}y\,dx\,dy.$$

Chapter 12 Review Exercises

3. Substituting $u(x,t) = v(x,t) + \psi(x)$ into the partial differential equation we obtain

$$k\frac{\partial^2 v}{\partial x^2} + k\psi''(x) = \frac{\partial v}{\partial t}.$$

This equation will be homogeneous provided ψ satisfies

$$k\psi'' = 0 \quad \text{or} \quad \psi = c_1 x + c_2.$$

Considering

$$u(0,t) = v(0,t) + \psi(0) = u_0$$

we set $\psi(0) = u_0$ so that $\psi(x) = c_1 x + u_0$. Now

$$-\frac{\partial u}{\partial x}\bigg|_{x=\pi} = -\frac{\partial v}{\partial x}\bigg|_{x=\pi} - \psi'(x) = v(\pi,t) + \psi(\pi) - u_1$$

is equivalent to

$$\frac{\partial v}{\partial x}\bigg|_{x=\pi} + v(\pi,t) = u_1 - \psi'(x) - \psi(\pi) = u_1 - c_1 - (c_1\pi + u_0),$$

which will be homogeneous when

$$u_1 - c_1 - c_1\pi - u_0 = 0 \quad \text{or} \quad c_1 = \frac{u_1 - u_0}{1+\pi}.$$

The steady-state solution is

$$\psi(x) = \left(\frac{u_1 - u_0}{1 + \pi}\right) x + u_0.$$

6. The boundary-value problem is

$$\frac{\partial^2 u}{\partial x^2} + x^2 = \frac{\partial^2 u}{\partial t^2}, \quad 0 < x < 1, \quad t > 0,$$

$$u(0, t) = 1, \quad u(1, t) = 0, \quad t > 0,$$

$$u(x, 0) = f(x), \quad u_t(x, 0) = 0, \quad 0 < x < 1.$$

Substituting $u(x, t) = v(x, t) + \psi(x)$ into the partial differential equation gives

$$\frac{\partial^2 v}{\partial x^2} + \psi''(x) + x^2 = \frac{\partial^2 v}{\partial t^2}.$$

This equation will be homogeneous provided $\psi''(x) + x^2 = 0$ or

$$\psi(x) = -\frac{1}{12} x^4 + c_1 x + c_2.$$

From $\psi(0) = 1$ and $\psi(1) = 0$ we obtain $c_1 = -11/12$ and $c_2 = 1$. The new problem is

$$\frac{\partial^2 v}{\partial x^2} = \frac{\partial^2 v}{\partial t^2}, \quad 0 < x < 1, \quad t > 0,$$

$$v(0, t) = 0, \quad v(1, t) = 0, \quad t > 0,$$

$$v(x, 0) = f(x) - \psi(x), \quad v_t(x, 0) = 0, \quad 0 < x < 1.$$

From Section 12.4 in the text we see that $B_n = 0$,

$$A_n = 2 \int_0^1 [f(x) - \psi(x)] \sin n\pi x \, dx = 2 \int_0^1 \left[f(x) + \frac{1}{12} x^4 + \frac{11}{12} x - 1 \right] \sin n\pi x \, dx,$$

and

$$v(x, t) = \sum_{n=1}^{\infty} A_n \cos n\pi t \sin n\pi x.$$

Thus

$$u(x, t) = v(x, t) + \psi(x) = -\frac{1}{12} x^4 - \frac{11}{12} x + 1 + \sum_{n=1}^{\infty} A_n \cos n\pi t \sin n\pi x.$$

9. Using $u = XY$ and λ^2 as a separation constant leads to

$$X'' - \lambda^2 X = 0,$$

and

$$Y'' + \lambda^2 Y = 0,$$

$$Y(0) = 0,$$

$$Y(\pi) = 0.$$

171

Then

$$Y = c_1 \sin ny \quad \text{and} \quad X = c_2 e^{-nx}$$

for $n = 1, 2, 3, \ldots$ (since u must be bounded as $x \to \infty$) so that

$$u = \sum_{n=1}^{\infty} A_n e^{-nx} \sin ny.$$

Imposing

$$u(0, y) = 50 = \sum_{n=1}^{\infty} A_n \sin ny$$

gives

$$A_n = \frac{2}{\pi} \int_0^{\pi} 50 \sin ny \, dy = \frac{100}{n\pi}[1 - (-1)^n]$$

so that

$$u(x, y) = \sum_{n=1}^{\infty} \frac{100}{n\pi}[1 - (-1)^n]e^{-nx} \sin ny.$$

12. Substituting $u(x, t) = v(x, t) + \psi(x)$ into the partial differential equation gives

$$k \frac{\partial^2 v}{\partial x^2} + k\psi'' + \sin 2\pi x = \frac{\partial v}{\partial t}.$$

This equation will be homogeneous provided ψ satisfies

$$k\psi'' + \sin 2\pi x = 0.$$

The general solution of this equation is

$$\psi(x) = \frac{1}{4k\pi^2} \sin 2\pi x + c_1 x + c_2.$$

From $\psi(0) = \psi(1) = 0$ we find that $c_1 = c_2 = 0$ and

$$\psi(x) = \frac{1}{4k\pi^2} \sin 2\pi x.$$

Now the new problem is

$$k \frac{\partial^2 v}{\partial x^2} = \frac{\partial v}{\partial t}, \quad 0 < x < 1, \quad t > 0$$

$$v(0, t) = 0, \quad v(1, t) = 0, \quad t > 0$$

$$v(x, 0) = \sin \pi x - \psi(x), \quad 0 < x < 1.$$

If we let $v = XT$ then

$$\frac{X''}{X} = \frac{T'}{kT} = -\lambda^2$$

gives the separated differential equations

$$X'' + \lambda^2 X = 0 \quad \text{and} \quad T' + k\lambda^2 T = 0.$$

The respective solutions are

$$X(x) = c_3 \cos \lambda x + c_4 \sin \lambda x$$

$$T(t) = c_5 e^{-k\lambda^2 t}.$$

From $X(0) = 0$ we get $c_3 = 0$ and from $X(1) = 0$ we find $\lambda = n\pi$ for $n = 1, 2, 3, \ldots$. Consequently, it follows that

$$v(x,t) = \sum_{n=1}^{\infty} A_n e^{-kn^2\pi^2 t} \sin n\pi x$$

where

$$v(x,0) = \sin \pi x - \frac{1}{4k\pi^2} \sin 2\pi x = 0$$

implies

$$A_n = 2 \int_0^1 \left(\sin \pi x - \frac{1}{4k\pi^2} \sin 2\pi x \right) \sin n\pi x \, dx.$$

By orthogonality $A_n = 0$ for $n = 3, 4, 5, \ldots$, and only A_1 and A_2 can be nonzero. We have

$$A_1 = 2 \left[\int_0^1 \sin^2 \pi x \, dx - \frac{1}{4k\pi^2} \int_0^1 \sin 2\pi x \sin \pi x \, dx \right] = 2 \int_0^1 \frac{1}{2}(1 - \cos 2\pi x) \, dx = 1$$

and

$$A_2 = 2 \left[\int_0^1 \sin \pi x \sin 2\pi x \, dx - \frac{1}{4k\pi^2} \int_0^1 \sin^2 2\pi x \, dx \right]$$

$$= -\frac{1}{2k\pi^2} \int_0^1 \frac{1}{2}(1 - \cos 4\pi x) \, dx = -\frac{1}{4k\pi^2}.$$

Therefore

$$v(x,t) = A_1 e^{-k\pi^2 t} \sin \pi x + A_2 e^{-k4\pi^2 t} \sin 2\pi x$$

$$= e^{-k\pi^2 t} \sin \pi x - \frac{1}{4k\pi^2} e^{-4k\pi^2 t} \sin 2\pi x$$

and

$$u(x,t) = v(x,t) + \psi(x) = e^{-k\pi^2 t} \sin \pi x + \frac{1}{4k\pi^2}(1 - e^{-4k\pi^2 t}) \sin 2\pi x.$$

13 Boundary-Value Problems in Other Coordinate Systems

3. We have

$$A_0 = \frac{1}{2\pi} \int_0^{2\pi} (2\pi\theta - \theta^2) \, d\theta = \frac{2\pi^2}{3}$$

$$A_n = \frac{1}{\pi} \int_0^{2\pi} (2\pi\theta - \theta^2) \cos n\theta \, d\theta = -\frac{4}{n^2}$$

$$B_n = \frac{1}{\pi} \int_0^{2\pi} (2\pi\theta - \theta^2) \sin n\theta \, d\theta = 0$$

and so

$$u(r, \theta) = \frac{2\pi^2}{3} - 4 \sum_{n=1}^{\infty} \frac{r^n}{n^2} \cos n\theta.$$

6. We solve

$$\frac{\partial^2 u}{\partial r^2} + \frac{1}{r} \frac{\partial u}{\partial r} + \frac{1}{r^2} \frac{\partial^2 u}{\partial \theta^2} = 0, \quad 0 < \theta < \frac{\pi}{2}, \quad 0 < r < c,$$

$$u(c, \theta) = f(\theta), \quad 0 < \theta < \frac{\pi}{2},$$

$$u(r, 0) = 0, \quad u(r, \pi/2) = 0, \quad 0 < r < c.$$

Proceeding as in Example 1 in the text we obtain the separated differential equations

$$r^2 R'' + rR' - \lambda^2 R = 0$$

$$\Theta'' + \lambda^2 \Theta = 0$$

with solutions

$$\Theta(\theta) = c_1 \cos \lambda\theta + c_2 \sin \lambda\theta$$

$$R(r) = c_3 r^\lambda + c_4 r^{-\lambda}.$$

Since we want $R(r)$ to be bounded as $r \to 0$ we require $c_4 = 0$. Applying the boundary conditions $\Theta(0) = 0$ and $\Theta(\pi/2) = 0$ we find that $c_1 = 0$ and $\lambda = 2n$ for $n = 1, 2, 3, \ldots$. Therefore

$$u(r, \theta) = \sum_{n=1}^{\infty} A_n r^{2n} \sin 2n\theta.$$

From

$$u(c, \theta) = f(\theta) = \sum_{n=1}^{\infty} A_n c^n \sin 2n\theta$$

we find

$$A_n = \frac{4}{\pi c^{2n}} \int_0^{\pi/2} f(\theta) \sin 2n\theta \, d\theta.$$

9. Proceeding as in Example 1 in the text and again using the periodicity of $u(r, \theta)$, we have

$$\Theta(\theta) = c_1 \cos \lambda\theta + c_2 \sin \lambda\theta$$

where $\lambda = n$ for $n = 0, 1, 2, \ldots$. Then

$$R(r) = c_3 r^n + c_4 r^{-n}.$$

[We do not have $c_4 = 0$ in this case since $0 < a \le r$.] Since $u(b, \theta) = 0$ we have

$$u(r, \theta) = A_0 \ln \frac{r}{b} + \sum_{n=1}^{\infty} \left[\left(\frac{b}{r}\right)^n - \left(\frac{r}{b}\right)^n \right] [A_n \cos n\theta + B_n \sin n\theta].$$

From

$$u(a, \theta) = f(\theta) = A_0 \ln \frac{a}{b} + \sum_{n=1}^{\infty} \left[\left(\frac{b}{a}\right)^n - \left(\frac{a}{b}\right)^n \right] [A_n \cos n\theta + B_n \sin n\theta]$$

we find

$$A_0 \ln \frac{a}{b} = \frac{1}{2\pi} \int_0^{2\pi} f(\theta) \, d\theta,$$

$$\left[\left(\frac{b}{a}\right)^n - \left(\frac{a}{b}\right)^n \right] A_n = \frac{1}{\pi} \int_0^{2\pi} f(\theta) \cos n\theta \, d\theta,$$

and

$$\left[\left(\frac{b}{a}\right)^n - \left(\frac{a}{b}\right)^n \right] B_n = \frac{1}{\pi} \int_0^{2\pi} f(\theta) \sin n\theta \, d\theta.$$

12. Letting $u(r, \theta) = v(r, \theta) + \psi(\theta)$ we obtain $\psi''(\theta) = 0$ and so $\psi(\theta) = c_1 \theta + c_2$. From $\psi(0) = 0$ and $\psi(\pi) = u_0$ we find, in turn, $c_2 = 0$ and $c_1 = u_0/\pi$. Therefore $\psi(\theta) = \frac{u_0}{\pi} \theta$. Now $u(1, \theta) = v(1, \theta) + \psi(\theta)$ so that $v(1, \theta) = u_0 - \frac{u_0}{\pi} \theta$. From

$$v(r, \theta) = \sum_{n=1}^{\infty} A_n r^n \sin n\theta \quad \text{and} \quad v(1, \theta) = \sum_{n=1}^{\infty} A_n \sin n\theta$$

we obtain

$$A_n = \frac{2}{\pi} \int_0^{\pi} \left(u_0 - \frac{u_0}{\pi} \theta \right) \sin n\theta \, d\theta = \frac{2u_0}{\pi n}.$$

Thus

$$u(r, \theta) = \frac{u_0}{\pi} \theta + \frac{2u_0}{\pi} \sum_{n=1}^{\infty} \frac{r^n}{n} \sin n\theta.$$

175

3. Referring to Example 2 in the text we have

$$R(r) = c_1 J_0(\lambda r) + c_2 Y_0(\lambda r)$$

$$Z(z) = c_3 \cosh \lambda z + c_4 \sinh \lambda z$$

where $c_2 = 0$ and $J_0(2\lambda) = 0$ defines the positive eigenvalues λ_n. From $Z(4) = 0$ we obtain

$$c_3 \cosh 4\lambda_n + c_4 \sinh 4\lambda_n = 0 \quad \text{or} \quad c_4 = -c_3 \frac{\cosh 4\lambda_n}{\sinh 4\lambda_n}.$$

Then

$$Z(z) = c_3 \left[\cosh \lambda_n z - \frac{\cosh 4\lambda_n}{\sinh 4\lambda_n} \sinh \lambda_n z \right] = c_3 \frac{\sinh 4\lambda_n \cosh \lambda_n z - \cosh 4\lambda_n \sinh \lambda_n z}{\sinh 4\lambda_n}$$

$$= c_3 \frac{\sinh \lambda_n(4 - z)}{\sinh 4\lambda_n}$$

and

$$u(r, z) = \sum_{n=1}^{\infty} A_n \frac{\sinh \lambda_n(4 - z)}{\sinh 4\lambda_n} J_0(\lambda_n r).$$

From

$$u(r, 0) = u_0 = \sum_{n=1}^{\infty} A_n J_0(\lambda_n r)$$

we obtain

$$A_n = \frac{2u_0}{4J_1^2(2\lambda_n)} \int_0^2 r J_0(\lambda_n r)\, dr = \frac{u_0}{\lambda_n J_1(2\lambda_n)}.$$

Thus the temperature in the cylinder is

$$u(r, z) = u_0 \sum_{n=1}^{\infty} \frac{\sinh \lambda_n(4 - z) J_0(\lambda_n r)}{\lambda_n \sinh 4\lambda_n J_1(2\lambda_n)}.$$

6. If the edge $r = c$ is insulated we have the boundary condition $u_r(c, t) = 0$. Letting $u(r, t) = R(r)T(t)$ and separating variables we obtain

$$\frac{R'' + \frac{1}{r} R'}{R} = \frac{T'}{kT} = \mu \quad \text{and} \quad R'' + \frac{1}{r} R' - \mu R = 0, \quad T' - \mu k T = 0.$$

From the second equation we find $T(t) = e^{\mu k t}$. If $\mu > 0$, $T(t)$ increases without bound as $t \to \infty$. Thus we assume $\mu = -\lambda^2 \le 0$. Now

$$R'' + \frac{1}{r} R' + \lambda^2 R = 0$$

is a parametric Bessel equation with solution

$$R(r) = c_1 J_0(\lambda r) + c_2 Y_0(\lambda r).$$

176

Since Y_0 is unbounded as $r \to 0$ we take $c_2 = 0$. Then $R(r) = c_1 J_0(\lambda r)$ and the boundary condition $u_r(c, t) = R'(c)T(t) = 0$ implies

$$R'(c) = \lambda c_1 J_0'(\lambda c) = 0.$$

This defines an eigenvalue $\lambda = 0$ and positive eigenvalues λ_n. Thus

$$u(r, t) = A_0 + \sum_{n=1}^{\infty} A_n J_0(\lambda_n r) e^{-\lambda_n^2 kt}.$$

From

$$u(r, 0) = f(r) = A_0 + \sum_{n=1}^{\infty} A_n J_0(\lambda_n r)$$

we find

$$A_0 = \frac{2}{c^2} \int_0^c r f(r)\, dr$$

$$A_n = \frac{2}{c^2 J_0^2(\lambda_n c)} \int_0^c r J_0(\lambda_n r) f(r)\, dr.$$

9. Substituting $u(r, t) = v(r, t) + \psi(r)$ into the partial differential equation gives

$$\frac{\partial^2 v}{\partial r^2} + \frac{1}{r}\frac{\partial v}{\partial r} + \psi'' + \frac{1}{r}\psi' = \frac{\partial v}{\partial t}.$$

This equation will be homogeneous provided $\psi'' + \frac{1}{r}\psi' = 0$ or

$$\psi(r) = c_1 \ln r + c_2.$$

Since $\ln r$ is unbounded as $r \to 0$ we take $c_1 = 0$. Then $\psi(r) = c_2$ and using

$$u(2, t) = v(2, t) + \psi(2) = 100$$

we set $c_2 = \psi(r) = 100$. Referring to Problem 6 above, the solution of

$$\frac{\partial^2 v}{\partial r^2} + \frac{1}{r}\frac{\partial v}{\partial r} = \frac{\partial v}{\partial t}, \quad 0 < r < 2, \quad t > 0$$

is

$$v(r, t) = c_1 J_0(\lambda r) e^{\mu t}.$$

The boundary conditions

$$v(2, t) = 0, \quad t > 0,$$

$$v(r, 0) = u(r, 0) - \psi(r)$$

then give

$$v(r, t) = \sum_{n=1}^{\infty} A_n J_0(\lambda_n r) e^{-\lambda_n^2 t}$$

where

$$A_n = \frac{2}{2^2 J_1^2(2\lambda_n)} \int_0^2 r J_0(\lambda_n r)[u(r,0) - \psi(r)]\, dr$$

$$= \frac{1}{2 J_1^2(2\lambda_n)} \left[\int_0^1 r J_0(\lambda_n r)[200 - 100]\, dr + \int_1^2 r J_0(\lambda_n r)[100 - 100]\, dr \right]$$

$$= \frac{50}{J_1^2(2\lambda_n)} \int_0^1 r J_0(\lambda_n r)\, dr \qquad \boxed{x = \lambda_n r, \ dx = \lambda_n\, dr}$$

$$= \frac{50}{J_1^2(2\lambda_n)} \int_0^{\lambda_n} \frac{1}{\lambda_n^2} x J_0(x)\, dx$$

$$= \frac{50}{\lambda_n^2 J_1^2(2\lambda_n)} \int_0^{\lambda_n} \frac{d}{dx}[x J_1(x)]\, dx \qquad \boxed{\text{see (4) of Section 11.5 in text}}$$

$$= \frac{50}{\lambda_n^2 J_1^2(2\lambda_n)} (x J_1(x)) \Big|_0^{\lambda_n} = \frac{50 J_1(\lambda_n)}{\lambda_n J_1^2(2\lambda_n)}.$$

Thus

$$u(r,t) = v(r,t) + \psi(r) = 100 + 50 \sum_{n=1}^{\infty} \frac{J_1(\lambda_n) J_0(\lambda_n r)}{\lambda_n J_1^2(2\lambda_n)} e^{-\lambda^2 t}.$$

12. (a) First we see that

$$\frac{R''\Theta + \dfrac{1}{r} R'\Theta + \dfrac{1}{r^2} R\Theta''}{R\Theta} = \frac{T''}{a^2 T} = -\lambda^2.$$

This gives $T'' + a^2 \lambda^2 T = 0$. Then from

$$\frac{R'' + \dfrac{1}{r} R' + \lambda^2 R}{-R/r^2} = \frac{\Theta''}{\Theta} = -\nu^2$$

we get $\Theta'' + \nu^2 \Theta = 0$ and $r^2 R'' + r R' + (\lambda^2 r^2 - \nu^2) R = 0$.

(b) The general solutions of the differential equations in part (a) are

$$T = c_1 \cos a\lambda t + c_2 \sin a\lambda t$$

$$\Theta = c_3 \cos \nu\theta + c_4 \cos \nu\theta$$

$$R = c_5 J_\nu(\lambda r) + c_6 Y_\nu(\lambda r).$$

(c) Implicitly we expect $u(r,\theta,t) = u(r,\theta + 2\pi, t)$ and so Θ must be 2π-periodic. Therefore $\nu = n$, $n = 0, 1, 2, \ldots$. The corresponding eigenfunctions are 1, $\cos\theta$, $\cos 2\theta$, $\ldots$, $\sin\theta$, $\sin 2\theta$, $\ldots$. Arguing that $u(r,\theta,t)$ is bounded as $r \to 0$ we then define $c_6 = 0$ and so $R = c_3 J_n(\lambda r)$. But $R(c) = 0$ gives $J_n(\lambda c) = 0$; this equation defines the eigenvalues λ_n. For each n, $\lambda_{ni} = x_{ni}/c$, $i = 1, 2, 3, \ldots$.

(d) $u(r,\theta,t) = \sum_{i=1}^{n}(A_{0i}\cos a\lambda_{0i}t + B_{0i}\sin a\lambda_{0i}t)J_0(\lambda_{0i}r)$

$$+ \sum_{n=1}^{\infty}\sum_{i=1}^{\infty}\Big[(A_{ni}\cos a\lambda_{ni}t + B_{ni}\sin a\lambda_{ni}t)\cos n\theta$$

$$+ (C_{ni}\cos a\lambda_{ni}t + D_{ni}\sin a\lambda_{ni}t)\sin n\theta\Big]J_n(\lambda_{ni}r)$$

Exercises 13.3

3. The coefficients are given by

$$A_n = \frac{2n+1}{2c^n}\int_0^{\pi}\cos\theta P_n(\cos\theta)\sin\theta\,d\theta = \frac{2n+1}{2c^n}\int_0^{\pi}P_1(\cos\theta)P_n(\cos\theta)\sin\theta\,d\theta$$

$$\boxed{x = \cos\theta,\ dx = -\sin\theta\,d\theta}$$

$$= \frac{2n+1}{2c^n}\int_{-1}^{1}P_1(x)P_n(x)\,dx.$$

Since $P_n(x)$ and $P_m(x)$ are orthogonal for $m \neq n$, $A_n = 0$ for $n \neq 1$ and

$$A_1 = \frac{2(1)+1}{2c^1}\int_{-1}^{1}P_1(x)P_1(x)\,dx = \frac{3}{2c}\int_{-1}^{1}x^2dx = \frac{1}{c}.$$

Thus

$$u(r,\theta) = \frac{r}{c}P_1(\cos\theta) = \frac{r}{c}\cos\theta.$$

6. Referring to Example 1 in the text we have

$$R(r) = c_1 r^n \quad\text{and}\quad \Theta(\theta) = P_n(\cos\theta).$$

Now $\Theta(\pi/2) = 0$ implies that n is odd, so

$$u(r,\theta) = \sum_{n=0}^{\infty}A_{2n+1}r^{2n+1}P_{2n+1}(\cos\theta).$$

From

$$u(c,\theta) = f(\theta) = \sum_{n=0}^{\infty}A_{2n+1}c^{2n+1}P_{2n+1}(\cos\theta)$$

we see that

$$A_{2n+1}c^{2n+1} = (4n+3)\int_0^{\pi/2}f(\theta)\sin\theta\,P_{2n+1}(\cos\theta)\,d\theta.$$

Thus

$$u(r,\theta) = \sum_{n=0}^{\infty}A_{2n+1}r^{2n+1}P_{2n+1}(\cos\theta)$$

179

where

$$A_{2n+1} = \frac{4n+3}{c^{2n+1}} \int_0^{\pi/2} f(\theta) \sin\theta \, P_{2n+1}(\cos\theta) \, d\theta.$$

9. Checking the hint, we find

$$\frac{1}{r}\frac{\partial^2}{\partial r^2}(ru) = \frac{1}{r}\frac{\partial}{\partial r}\left[r\frac{\partial u}{\partial r} + u\right] = \frac{1}{r}\left[r\frac{\partial^2 u}{\partial r^2} + \frac{\partial u}{\partial r} + \frac{\partial u}{\partial r}\right] = \frac{\partial^2 u}{\partial r^2} + \frac{2}{r}\frac{\partial u}{\partial r}.$$

The partial differential equation then becomes

$$\frac{\partial^2}{\partial r^2}(ru) = r\frac{\partial u}{\partial t}.$$

Now, letting $ru(r,t) = v(r,t) + \psi(r)$, since the boundary condition is nonhomogeneous, we obtain

$$\frac{\partial^2}{\partial r^2}[v(r,t) + \psi(r)] = r\frac{\partial}{\partial t}\left[\frac{1}{r}v(r,t) + \psi(r)\right]$$

or

$$\frac{\partial^2 v}{\partial r^2} + \psi''(r) = \frac{\partial v}{\partial t}.$$

This differential equation will be homogeneous if $\psi''(r) = 0$ or $\psi(r) = c_1 r + c_2$. Now

$$u(r,t) = \frac{1}{r}v(r,t) + \frac{1}{r}\psi(r) \quad \text{and} \quad \frac{1}{r}\psi(r) = c_1 + \frac{c_2}{r}.$$

Since we want $u(r,t)$ to be bounded as r approaches 0, we require $c_2 = 0$. Then $\psi(r) = c_1 r$. When $r = 1$

$$u(1,t) = v(1,t) + \psi(1) = v(1,t) + c_1 = 100,$$

and we will have the homogeneous boundary condition $v(1,t) = 0$ when $c_1 = 100$. Consequently, $\psi(r) = 100r$. The initial condition

$$u(r,0) = \frac{1}{r}v(r,0) + \frac{1}{r}\psi(r) = \frac{1}{r}v(r,0) + 100 = 0$$

implies $v(r,0) = -100r$. We are thus led to solve the new boundary-value problem

$$\frac{\partial^2 v}{\partial r^2} = \frac{\partial v}{\partial t}, \quad 0 < r < 1, \quad t > 0,$$

$$v(1,t) = 0, \quad \lim_{r\to 0}\frac{1}{r}v(r,t) < \infty,$$

$$v(r,0) = -100r.$$

Letting $v(r,t) = R(r)T(t)$ and separating variables leads to

$$R'' + \lambda^2 R = 0 \quad \text{and} \quad T' + \lambda^2 T = 0$$

with solutions

$$R(r) = c_3 \cos\lambda r + c_4 \sin\lambda r \quad \text{and} \quad T(t) = c_5 e^{-\lambda^2 t}.$$

180

The boundary conditions are equivalent to $R(1) = 0$ and $\lim_{r \to 0} \frac{1}{r} R(r) < \infty$. Since

$$\lim_{r \to 0} \frac{1}{r} R(r) = \lim_{r \to 0} \frac{c_3 \cos \lambda r}{r} + \lim_{r \to 0} \frac{c_4 \sin \lambda r}{r} = \lim_{r \to 0} \frac{c_3 \cos \lambda r}{r} + c_4 \lambda < \infty$$

we must have $c_3 = 0$. Then $R(r) = c_4 \sin \lambda r$, and $R(1) = 0$ implies $\lambda = n\pi$ for $n = 1, 2, 3, \ldots$. Thus

$$v_n(r, t) = A_n e^{-n^2 \pi^2 t} \sin n\pi r$$

for $n = 1, 2, 3, \ldots$. Using the condition $\lim_{r \to 0} \frac{1}{r} R(r) < \infty$ it is easily shown that there are no eigenvalues for $\lambda = 0$, nor does setting the common constant to $+\lambda^2$ when separating variables lead to any solutions. Now, by the superposition principle,

$$v(r, t) = \sum_{n=1}^{\infty} A_n e^{-n^2 \pi^2 t} \sin n\pi r.$$

The initial condition $v(r, 0) = -100r$ implies

$$-100r = \sum_{n=1}^{\infty} A_n \sin n\pi r.$$

This is a Fourier sine series and so

$$A_n = 2 \int_0^1 (-100r \sin n\pi r) \, dr = -200 \left[-\frac{r}{n\pi} \cos n\pi r \Big|_0^1 + \int_0^1 \frac{1}{n\pi} \cos n\pi r \, dr \right]$$

$$= -200 \left[-\frac{\cos n\pi}{n\pi} + \frac{1}{n^2 \pi^2} \sin n\pi r \Big|_0^1 \right] = -200 \left[-\frac{(-1)^n}{n\pi} \right] = \frac{(-1)^n 200}{n\pi}.$$

A solution of the problem is thus

$$u(r, t) = \frac{1}{r} v(r, t) + \frac{1}{r} \psi(r) = \frac{1}{r} \sum_{n=1}^{\infty} (-1)^n \frac{20}{n\pi} e^{-n^2 \pi^2 t} \sin n\pi r + \frac{1}{r}(100r)$$

$$= \frac{200}{\pi r} \sum_{n=1}^{\infty} \frac{(-1)^n}{n} e^{-n^2 \pi^2 t} \sin n\pi r + 100.$$

12. Proceeding as in Example 1 we obtain

$$\Theta(\theta) = P_n(\cos \theta) \quad \text{and} \quad R(r) = c_1 r^n + c_2 r^{-(n+1)}$$

so that

$$u(r, \theta) = \sum_{n=0}^{\infty} (A_n r^n + B_n r^{-(n+1)}) P_n(\cos \theta).$$

To satisfy $\lim_{r \to \infty} u(r, \theta) = -Er \cos \theta$ we must have $A_n = 0$ for $n = 2, 3, 4, \ldots$. Then

$$\lim_{r \to \infty} u(r, \theta) = -Er \cos \theta = A_0 \cdot 1 + A_1 r \cos \theta,$$

181

so $A_0 = 0$ and $A_1 = -E$. Thus

$$u(r, \theta) = -Er \cos \theta + \sum_{n=0}^{\infty} B_n r^{-(n+1)} P_n(\cos \theta).$$

Now

$$u(c, \theta) = 0 = -Ec \cos \theta + \sum_{n=0}^{\infty} B_n c^{-(n+1)} P_n(\cos \theta)$$

so

$$\sum_{n=0}^{\infty} B_n c^{-(n+1)} P_n(\cos \theta) = Ec \cos \theta$$

and

$$B_n c^{-(n+1)} = \frac{2n+1}{2} \int_0^{\pi} Ec \cos \theta \, P_n(\cos \theta) \sin \theta \, d\theta.$$

Now $\cos \theta = P_1(\cos \theta)$ so, for $n \neq 1$,

$$\int_0^{\pi} \cos \theta \, P_n(\cos \theta) \sin \theta \, d\theta = 0$$

by orthogonality. Thus $B_n = 0$ for $n \neq 1$ and

$$B_1 = \frac{3}{2} Ec^3 \int_0^{\pi} \cos^2 \theta \sin \theta \, d\theta = Ec^3.$$

Therefore,

$$u(r, \theta) = -Er \cos \theta + Ec^3 r^{-2} \cos \theta.$$

——— Chapter 13 Review Exercises ———

3. The conditions $\Theta(0) = 0$ and $\Theta(\pi) = 0$ applied to $\Theta = c_1 \cos \lambda \theta + c_2 \sin \lambda \theta$ give $c_1 = 0$ and $\lambda = n$, $n = 1, 2, 3, \ldots$, respectively. Thus we have the Fourier sine-series coefficients

$$A_n = \frac{2}{\pi} \int_0^{\pi} u_0(\pi \theta - \theta^2) \sin n\theta \, d\theta = \frac{4u_0}{n^3 \pi} [1 - (-1)^n].$$

Thus

$$u(r, \theta) = \frac{4u_0}{\pi} \sum_{n=1}^{\infty} \frac{1 - (-1)^n}{n^3} r^n \sin n\theta.$$

6. We solve

$$\frac{\partial^2 u}{\partial r^2} + \frac{1}{r} \frac{\partial u}{\partial r} + \frac{1}{r^2} \frac{\partial^2 u}{\partial \theta^2} = 0, \quad r > 1, \quad 0 < \theta < \pi,$$

$$u(r, 0) = 0, \quad u(r, \pi) = 0, \quad r > 1,$$

$$u(1, \theta) = f(\theta), \quad 0 < \theta < \pi.$$

Separating variables we obtain

$$\Theta(\theta) = c_1 \cos \lambda\theta + c_2 \sin \lambda\theta$$

$$R(r) = c_3 r^\lambda + c_4 r^{-\lambda}.$$

Applying the boundary conditions $\Theta(0) = 0$, and $\Theta(\pi) = 0$ gives $c_1 = 0$ and $\lambda = n$ for $n = 1, 2, 3, \ldots$. Assuming $f(\theta)$ to be bounded, we expect the solution $u(r, \theta)$ to also be bounded as $r \to \infty$. This requires that $c_3 = 0$. Therefore

$$u(r, \theta) = \sum_{n=1}^{\infty} A_n r^{-n} \sin n\theta.$$

From

$$u(1, \theta) = f(\theta) = \sum_{n=1}^{\infty} A_n \sin n\theta$$

we obtain

$$A_n = \frac{2}{\pi} \int_0^{\pi} f(\theta) \sin n\theta \, d\theta.$$

9. Referring to Example 2 in Section 13.2 we have

$$R(r) = c_1 J_0(\lambda r) + c_2 Y_0(\lambda r)$$

$$Z(z) = c_3 \cosh \lambda z + c_4 \sinh \lambda z$$

where $c_2 = 0$ and $J_0(2\lambda) = 0$ defines the positive eigenvalues λ_n. From $Z'(0) = 0$ we obtain $c_4 = 0$. Then

$$u(r, z) = \sum_{n=1}^{\infty} A_n \cosh \lambda_n z J_0(\lambda_n r).$$

From

$$u(r, 4) = 50 = \sum_{n=1}^{\infty} A_n \cosh 4\lambda_n J_0(\lambda_n r)$$

we obtain (as in Example 1 of Section 13.1)

$$A_n \cosh 4\lambda_n = \frac{2(50)}{4 J_1^2(2\lambda_n)} \int_0^2 r J_0(\lambda_n r) \, dr = \frac{50}{\lambda_n J_1(2\lambda_n)}.$$

Thus the temperature in the cylinder is

$$u(r, z) = 50 \sum_{n=1}^{\infty} \frac{\cosh \lambda_n z J_0(\lambda_n r)}{\lambda_n \cosh 4\lambda_n J_1(2\lambda_n)}.$$

12. Since

$$\frac{1}{r} \frac{\partial^2}{\partial r^2}(ru) = \frac{1}{r} \frac{\partial}{\partial r}\left[r \frac{\partial u}{\partial r} + u\right] = \frac{1}{r}\left[r \frac{\partial^2 u}{\partial r^2} + \frac{\partial u}{\partial r} + \frac{\partial u}{\partial r}\right] = \frac{\partial^2 u}{\partial r^2} + \frac{2}{r} \frac{\partial u}{r}$$

the differential equation becomes

$$\frac{1}{r} \frac{\partial^2}{\partial r^2}(ru) = \frac{\partial^2 u}{\partial t^2} \quad \text{or} \quad \frac{\partial^2}{\partial r^2}(ru) = r \frac{\partial^2 u}{\partial t^2}.$$

Letting $v(r,t) = ru(r,t)$ we obtain the boundary-value problem

$$\frac{\partial^2 v}{\partial r^2} = \frac{\partial^2 v}{\partial t^2}, \quad 0 < r < 1, \quad t > 0$$

$$\frac{\partial v}{\partial r}\bigg|_{r=1} - v(1,t) = 0, \quad t > 0$$

$$v(r,0) = rf(r), \quad \frac{\partial v}{\partial t}\bigg|_{t=0} = rg(r), \quad 0 < r < 1.$$

If we separate variables using $v(r,t) = R(r)T(t)$ then we obtain

$$R(r) = c_1 \cos \lambda r + c_2 \sin \lambda r$$

$$T(t) = c_3 \cos \lambda t + c_4 \sin \lambda t.$$

Since $u(r,t) = v(r,t)/r$, in order to insure boundedness at $r = 0$ we define $c_1 = 0$. Then $R(r) = c_2 \sin \lambda r$. Now the boundary condition $R'(1) - R(1) = 0$ implies $\lambda \cos \lambda - \sin \lambda = 0$. Thus, the eigenvalues λ_n are the positive solutions of $\tan \lambda = \lambda$. We now have

$$v_n(r,t) = (A_n \cos \lambda_n t + B_n \sin \lambda_n t) \sin \lambda_n r.$$

For the eigenvalue $\lambda = 0$,

$$R(r) = c_1 r + c_2 \quad \text{and} \quad T(t) = c_3 t + c_4,$$

and boundedness at $r = 0$ implies $c_2 = 0$. We then take

$$v_0(r,t) = A_0 tr + B_0 r$$

so that

$$v(r,t) = A_0 tr + B_0 r + \sum_{n=1}^{\infty} (a_n \cos \lambda_n t + B_n \sin \lambda_n t) \sin \lambda_n r.$$

Now

$$v(r,0) = rf(r) = B_0 r + \sum_{n=1}^{\infty} A_n \sin \lambda_n r.$$

Since $\{r, \sin \lambda_n r\}$ is an orthogonal set on $[0,1]$,

$$\int_0^1 r \sin \lambda_n r \, dr = 0 \quad \text{and} \quad \int_0^1 \sin \lambda_n r \sin \lambda_n r \, dr = 0$$

for $m \neq n$. Therefore

$$\int_0^1 r^2 f(r) \, dr = B_0 \int_0^1 r^2 \, dr = \frac{1}{3} B_0$$

and

$$B_0 = 3 \int_0^1 r^2 f(r) \, dr.$$

Also

$$\int_0^1 r f(r) \sin \lambda_n r \, dr = A_n \int_0^1 \sin^2 \lambda_n r \, dr$$

and

$$A_n = \frac{\int_0^1 r f(r) \sin \lambda_n r \, dr}{\int_0^1 \sin^2 \lambda_n r \, dr} .$$

Now

$$\int_0^1 \sin^2 \lambda_n r \, dr = \frac{1}{2} \int_0^1 (1 - \cos 2\lambda_n r) \, dr = \frac{1}{2} \left[1 - \frac{\sin 2\lambda_n}{2\lambda_n} \right] = \frac{1}{2} [1 - \cos^2 \lambda_n].$$

Since $\tan \lambda_n = \lambda_n$,

$$1 + \lambda_n^2 = 1 + \tan^2 \lambda_n = \sec^2 \lambda_n = \frac{1}{\cos^2 \lambda_n}$$

and

$$\cos^2 \lambda_n = \frac{1}{1 + \lambda_n^2} .$$

Then

$$\int_0^1 \sin^2 \lambda_n r \, dr = \frac{1}{2} \left[1 - \frac{1}{1 + \lambda_n^2} \right] = \frac{\lambda_n^2}{2(1 + \lambda_n^2)}$$

and

$$A_n = \frac{2(1 + \lambda_n^2)}{\lambda_n^2} \int_0^1 r f(r) \sin \lambda_n r \, dr.$$

Similarly, setting

$$\frac{\partial v}{\partial t} \bigg|_{t=0} = rg(r) = A_0 r + \sum_{n=1}^{\infty} B_n \lambda_n \sin \lambda_n r$$

we obtain

$$A_0 = 3 \int_0^1 r^2 g(r) \, dr$$

and

$$B_n = \frac{2(1 + \lambda_n^2)}{\lambda_n^3} \int_0^1 r g(r) \sin \lambda_n r \, dr.$$

Therefore, since $v(r, t) = ru(r, t)$ we have

$$u(r, t) = A_0 t + B_0 + \sum_{n=1}^{\infty} (A_n \cos \lambda_n t + B_n \sin \lambda_n t) \frac{\sin \lambda_n r}{r},$$

where the λ_n are solutions of $\tan \lambda = \lambda$ and

$$A_0 = 3 \int_0^1 r^2 g(r)\, dr$$

$$B_0 = 3 \int_0^1 r^2 f(r)\, dr$$

$$A_n = \frac{2(1 + \lambda_n^2)}{\lambda_n^2} \int_0^1 r f(r) \sin \lambda_n r\, dr$$

$$B_n = \frac{2(1 + \lambda_n^2)}{\lambda_n^3} \int_0^1 r g(r) \sin \lambda_n r\, dr$$

for $n = 1, 2, 3, \ldots$.

14 Integral Transform Method

_____ Exercises 14.1 _____

3. By the first translation theorem,

$$\mathscr{L}\left\{e^t \operatorname{erf}(\sqrt{t}\,)\right\} = \mathscr{L}\left\{\operatorname{erf}(\sqrt{t}\,)\right\}\Big|_{s \to s-1} = \frac{1}{s\sqrt{s+1}}\Big|_{s \to s-1} = \frac{1}{\sqrt{s}\,(s-1)}\,.$$

6. We first compute

$$\frac{\sinh a\sqrt{s}}{s \sinh \sqrt{s}} = \frac{e^{a\sqrt{s}} - e^{-a\sqrt{s}}}{s(e^{\sqrt{s}} - e^{-\sqrt{s}})} = \frac{e^{(a-1)\sqrt{s}} - e^{-(a+1)\sqrt{s}}}{s(1 - e^{-2\sqrt{s}})}$$

$$= \frac{e^{(a-1)\sqrt{s}}}{s}\left[1 + e^{-2\sqrt{s}} + e^{-4\sqrt{s}} + \cdots\right] - \frac{e^{-(a+1)\sqrt{s}}}{s}\left[1 + e^{-2\sqrt{s}} + e^{-4\sqrt{s}} + \cdots\right]$$

$$= \left[\frac{e^{-(1-a)\sqrt{s}}}{s} + \frac{e^{-(3-a)\sqrt{s}}}{s} + \frac{e^{-(5-a)\sqrt{s}}}{s} + \cdots\right]$$

$$\qquad - \left[\frac{e^{-(1+a)\sqrt{s}}}{s} + \frac{e^{-(3+a)\sqrt{s}}}{s} + \frac{e^{-(5+a)\sqrt{s}}}{s} + \cdots\right]$$

$$= \sum_{n=0}^{\infty}\left[\frac{e^{-(2n+1-a)\sqrt{s}}}{s} - \frac{e^{-(2n+1+a)\sqrt{s}}}{s}\right].$$

Then

$$\mathscr{L}\left\{\frac{\sinh a\sqrt{s}}{s \sinh \sqrt{s}}\right\} = \sum_{n=0}^{\infty}\left[\mathscr{L}\left\{\frac{e^{-(2n+1-a)\sqrt{s}}}{s}\right\} - \mathscr{L}\left\{-\frac{e^{-(2n+1+a)\sqrt{s}}}{s}\right\}\right]$$

$$= \sum_{n=0}^{\infty}\left[\operatorname{erfc}\left(\frac{2n+1-a}{2\sqrt{t}}\right) - \operatorname{erfc}\left(\frac{2n+1+a}{2\sqrt{t}}\right)\right]$$

$$= \sum_{n=0}^{\infty}\left(\left[1 - \operatorname{erf}\left(\frac{2n+1-a}{2\sqrt{t}}\right)\right] - \left[1 - \operatorname{erf}\left(\frac{2n+1+a}{2\sqrt{t}}\right)\right]\right)$$

$$= \sum_{n=0}^{\infty}\left[\operatorname{erf}\left(\frac{2n+1+a}{2\sqrt{t}}\right) - \operatorname{erf}\left(\frac{2n+1-a}{2\sqrt{t}}\right)\right].$$

9. $\displaystyle\int_{a}^{b} e^{-u^2}\,du = \int_{a}^{0} e^{-u^2}\,du + \int_{0}^{b} e^{-u^2}\,du = \int_{0}^{b} e^{-u^2}\,du - \int_{0}^{a} e^{-u^2}\,du$

$$= \frac{\sqrt{\pi}}{2}\operatorname{erf}(b) - \frac{\sqrt{\pi}}{2}\operatorname{erf}(a) = \frac{\sqrt{\pi}}{2}[\operatorname{erf}(b) - \operatorname{erf}(a)]$$

_____ **Exercises 14.2** _____

3. The solution of

$$a^2 \frac{d^2U}{dx^2} - s^2U = 0$$

is in this case

$$U(x, s) = c_1 e^{-(x/a)s} + c_2 e^{(x/a)s}.$$

Since $\lim_{x \to \infty} u(x, t) = 0$ we have $\lim_{x \to \infty} U(x, s) = 0$. Thus $c_2 = 0$ and

$$U(x, s) = c_1 e^{-(x/a)s}.$$

If $\mathscr{L}\{u(0, t)\} = \mathscr{L}\{f(t)\} = F(s)$ then $U(0, s) = F(s)$. From this we have $c_1 = F(s)$ and

$$U(x, s) = F(s)e^{-(x/a)s}.$$

Hence, by the second translation theorem,

$$u(x, t) = f\left(t - \frac{x}{a}\right)\mathscr{U}\left(t - \frac{x}{a}\right).$$

6. Transforming the partial differential equation gives

$$\frac{d^2U}{dx^2} - s^2U = -\frac{\omega}{s^2 + \omega^2} \sin \pi x.$$

Using undetermined coefficients we obtain

$$U(x, s) = c_1 \cosh sx + c_2 \sinh sx + \frac{\omega}{(s^2 + \pi^2)(s^2 + \omega^2)} \sin \pi x.$$

The transformed boundary conditions $U(0, s) = 0$ and $U(1, s) = 0$ give, in turn, $c_1 = 0$ and $c_2 = 0$. Therefore

$$U(x, s) = \frac{\omega}{(s^2 + \pi^2)(s^2 + \omega^2)} \sin \pi x$$

and

$$u(x, t) = \omega \sin \pi x \, \mathscr{L}^{-1}\left\{\frac{1}{(s^2 + \pi^2)(s^2 + \omega^2)}\right\}$$

$$= \frac{\omega}{\omega^2 - \pi^2} \sin \pi x \, \mathscr{L}^{-1}\left\{\frac{1}{\pi} \frac{\pi}{s^2 + \pi^2} - \frac{1}{\omega} \frac{\omega}{s^2 + \omega^2}\right\}$$

$$= \frac{\omega}{\pi(\omega^2 - \pi^2)} \sin \pi t \sin \pi x - \frac{1}{\omega^2 - \pi^2} \sin \omega t \sin \pi x.$$

9. Transforming the partial differential equation gives

$$\frac{d^2U}{dx^2} - s^2U = -sxe^{-x}.$$

Using undetermined coefficients we obtain

$$U(x, s) = c_1 e^{-sx} + c_2 e^{sx} - \frac{2s}{(s^2 - 1)^2} e^{-x} + \frac{s}{s^2 - 1} x e^{-x}.$$

The transformed boundary conditions $\lim_{x \to \infty} U(x, s) = 0$ and $U(0, s) = 0$ give, in turn, $c_2 = 0$ and $c_1 = 2s/(s^2 - 1)^2$. Therefore

$$U(x, s) = \frac{2s}{(s^2 - 1)^2} e^{-sx} - \frac{2s}{(s^2 - 1)^2} e^{-x} + \frac{s}{s^2 - 1} x e^{-x}.$$

From entries (13) and (26) in the Table of Laplace transforms we obtain

$$u(x, t) = \mathscr{L}^{-1} \left\{ \frac{2s}{(s^2 - 1)^2} e^{-sx} - \frac{2s}{(s^2 - 1)^2} e^{-x} + \frac{s}{s^2 - 1} x e^{-x} \right\}$$

$$= 2(t - x) \sinh(t - x) \,\mathscr{U}(t - x) - t e^{-x} \sinh t + x e^{-x} \cosh t.$$

12. (a) Transforming the partial differential equation and using the initial condition gives

$$k \frac{d^2 U}{dx^2} - sU = 0.$$

Since the domain of the variable x is an infinite interval we write the general solution of this differential equation as

$$U(x, s) = c_1 e^{-\sqrt{s/k}\, x} + c_2 e^{-\sqrt{s/k}\, x}.$$

Transforming the boundary conditions gives $U'(0, s) = -A/s$ and $\lim_{x \to \infty} U(x, s) = 0$. Hence we find $c_2 = 0$ and $c_1 = A\sqrt{k}/s\sqrt{s}$. From

$$U(x, s) = A\sqrt{k} \, \frac{e^{-\sqrt{s/k}\, x}}{s\sqrt{s}}$$

we see that

$$u(x, t) = A\sqrt{k} \, \mathscr{L}^{-1} \left\{ \frac{e^{-\sqrt{s/k}\, x}}{s\sqrt{s}} \right\}.$$

With the identification $a = x/\sqrt{k}$ it follows from (49) in the Table of Laplace transforms that

$$u(x, t) = A\sqrt{k} \left\{ 2\sqrt{\frac{t}{\pi}} e^{-x^2/4kt} - \frac{x}{\sqrt{k}} \left(x/2\sqrt{kt} \right) \right\}$$

$$= 2A\sqrt{\frac{kt}{\pi}} e^{-x^2/4kt} - Ax \, \mathrm{erfc}\left(x/2\sqrt{kt} \right).$$

Since $\mathrm{erfc}(0) = 1$,

$$\lim_{t \to \infty} u(x, t) = \lim_{t \to \infty} \left(2A\sqrt{\frac{kt}{\pi}} e^{-x^2/4kt} - Ax \, \mathrm{erfc}\, \frac{x}{2\sqrt{kt}} \right) = \infty.$$

189

(b)

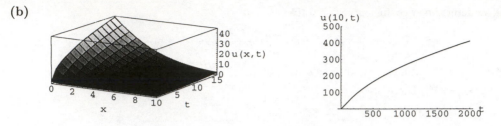

15. We use

$$U(x,s) = c_1 e^{-\sqrt{s}\,x} + c_2 e^{\sqrt{s}\,x} + \frac{u_0}{s}.$$

The condition $\lim_{x\to\infty} u(x,t) = u_0$ implies $\lim_{x\to\infty} U(x,s) = u_0/s$, so we define $c_2 = 0$. Then

$$U(x,s) = c_1 e^{-\sqrt{s}\,x} + \frac{u_0}{s}.$$

The transform of the remaining boundary conditions gives

$$\frac{dU}{dx}\bigg|_{x=0} = U(0,s).$$

This condition yields $c_1 = -u_0/s(\sqrt{s}+1)$. Thus

$$U(x,s) = -u_0\,\frac{e^{-\sqrt{s}\,x}}{s(\sqrt{s}+1)} + \frac{u_0}{s}$$

and

$$u(x,t) = -u_0\,\mathcal{L}^{-1}\left\{\frac{e^{-x\sqrt{s}}}{s(\sqrt{s}+1)}\right\} + u_0\,\mathcal{L}^{-1}\left\{\frac{1}{s}\right\}$$

$$= u_0 e^{x+t}\,\mathrm{erfc}\left(\sqrt{t} + \frac{x}{2\sqrt{t}}\right) - u_0\,\mathrm{erfc}\left(\frac{x}{2\sqrt{t}}\right) + u_0 \qquad \boxed{\text{By entry (5) in Table 14.1}}$$

18. We use

$$U(x,s) = c_1 e^{-\sqrt{s}\,x} + c_2 e^{\sqrt{s}\,x}.$$

The condition $\lim_{x\to\infty} u(x,t) = 0$ implies $\lim_{x\to\infty} U(x,s) = 0$, so we define $c_2 = 0$. Then $U(x,s) = c_1 e^{-\sqrt{s}\,x}$. The transform of the remaining boundary condition gives

$$\frac{dU}{dx}\bigg|_{x=0} = -F(s)$$

where $F(s) = \mathcal{L}\{f(t)\}$. This condition yields $c_1 = F(s)/\sqrt{s}$. Thus

$$U(x,s) = F(s)\,\frac{e^{-\sqrt{s}\,x}}{\sqrt{s}}.$$

Using entry (44) in the Table of Laplace transforms and the convolution theorem we obtain

$$u(x,t) = \mathcal{L}^{-1}\left\{F(s)\cdot\frac{e^{-\sqrt{s}\,x}}{\sqrt{s}}\right\} = \frac{1}{\sqrt{\pi}}\int_0^t f(\tau)\,\frac{e^{-x^2/4(t-\tau)}}{\sqrt{t-\tau}}\,d\tau.$$

21. Transforming the partial differential equation gives

$$\frac{d^2U}{dx^2} - sU = 0$$

and so

$$U(x, s) = c_1 e^{-\sqrt{s}\, x} + c_2 e^{\sqrt{s}\, x}.$$

The condition $\lim_{x \to -\infty} u(x, t) = 0$ implies $\lim_{x \to -\infty} U(x, s) = 0$, so we define $c_1 = 0$. The transform of the remaining boundary condition gives

$$\frac{dU}{dx}\bigg|_{x=1} = \frac{100}{s} - U(1, s).$$

This condition yields

$$c_2 \sqrt{s}\, e^{\sqrt{s}} = \frac{100}{s} - c_2 e^{\sqrt{s}}$$

from which it follows that

$$c_2 = \frac{100}{s(\sqrt{s} + 1)} e^{-\sqrt{s}}.$$

Thus

$$U(x, s) = 100 \frac{e^{-(1-x)\sqrt{s}}}{s(\sqrt{s} + 1)}.$$

Using entry (49) in the Table of Laplace transforms we obtain

$$u(x, t) = 100 \mathcal{L}^{-1} \left\{ \frac{e^{-(1-x)\sqrt{s}}}{s(\sqrt{s} + 1)} \right\} = 100 \left[-e^{1-x+t} \operatorname{erfc}\left(\sqrt{t} + \frac{1-x}{\sqrt{t}}\right) + \operatorname{erfc}\left(\frac{1-x}{2\sqrt{t}}\right) \right].$$

24. The transform of the partial differential equation is

$$k\frac{d^2U}{dx^2} - hU + h\frac{u_m}{s} = sU - u_0$$

or

$$k\frac{d^2U}{dx^2} - (h + s)U = -h\frac{u_m}{s} - u_0.$$

By undetermined coefficients we find

$$U(x, s) = c_1 e^{\sqrt{(h+s)/k}\, x} + c_2 e^{-\sqrt{(h+s)/k}\, x} + \frac{hu_m + u_0 s}{s(s + h)}.$$

The transformed boundary conditions are $U'(0, s) = 0$ and $U'(L, s) = 0$. These conditions imply $c_1 = 0$ and $c_2 = 0$. By partial fractions we then get

$$U(x, s) = \frac{hu_m + u_0 s}{s(s + h)} = \frac{u_m}{s} - \frac{u_m}{s + h} + \frac{u_0}{s + h}.$$

Therefore,

$$u(x, t) = u_m \mathcal{L}^{-1}\left\{\frac{1}{s}\right\} - u_m \mathcal{L}^{-1}\left\{\frac{1}{s+h}\right\} + u_0 \mathcal{L}^{-1}\left\{\frac{1}{s+h}\right\} = u_m - u_m e^{-ht} + u_0 e^{-ht}.$$

27. We use

$$U(x,s) = c_1 e^{-\sqrt{RCs+RG}\,x} + c_2 e^{\sqrt{RCs+RG}} + \frac{Cu_0}{Cs+G}.$$

The condition $\lim_{x\to\infty} \partial u/\partial x = 0$ implies $\lim_{x\to\infty} dU/dx = 0$, so we define $c_2 = 0$. Applying $U(0,s) = 0$ to

$$U(x,s) = c_1 e^{-\sqrt{RCsRG}\,x} + \frac{Cu_0}{Cs+G}$$

gives $c_1 = -Cu_0/(Cs+G)$. Therefore

$$U(x,s) = -Cu_0 \frac{e^{-\sqrt{RCs+RG}\,x}}{Cs+G} + \frac{Cu_0}{Cs+G}$$

and

$$u(x,t) = u_0 \mathscr{L}^{-1}\left\{\frac{1}{s+G/C}\right\} - u_0 \mathscr{L}^{-1}\left\{\frac{e^{-x\sqrt{RC}\sqrt{s+G/C}}}{s+G/C}\right\}$$

$$= u_0 e^{-Gt/C} - u_0 e^{-Gt/C}\,\mathrm{erfc}\left(\frac{x\sqrt{RC}}{2\sqrt{t}}\right)$$

$$= u_0 e^{-Gt/C}\left[1 - \mathrm{erfc}\left(\frac{x}{2}\sqrt{\frac{RC}{t}}\right)\right]$$

$$= u_0 e^{-Gt/C}\,\mathrm{erf}\left(\frac{x}{2}\sqrt{\frac{RC}{t}}\right).$$

30. (a) We use

$$U(x,s) = c_1 e^{-(s/a)x} + c_2 e^{(s/a)x} + \frac{v_0^2 F_0}{(a^2 - v_0^2)s^2}\,e^{-(s/v_0)x}.$$

The condition $\lim_{x\to\infty} u(x,t) = 0$ implies $\lim_{x\to\infty} U(x,s) = 0$, so we must define $c_2 = 0$. Consequently

$$U(x,s) = c_1 e^{-(s/a)x} + \frac{v_0^2 F_0}{(a^2 - v_0^2)s^2}\,e^{-(s/v_0)x}.$$

The remaining boundary condition transforms into $U(0,s) = 0$. From this we find

$$c_1 = -v_0^2 F_0/(a^2 - v_0^2)s^2.$$

Therefore, by the second translation theorem

$$U(x,s) = -\frac{v_0^2 F_0}{(a^2 - v_0^2)s^2}\,e^{-(s/a)x} + \frac{v_0^2 F_0}{(a^2 - v_0^2)s^2}\,e^{-(s/v_0)x}$$

and

$$u(x,t) = \frac{v_0^2 F_0}{a^2 - v_0^2}\left[\mathscr{L}^{-1}\left\{\frac{e^{-(x/v_0)s}}{s^2}\right\} - \mathscr{L}^{-1}\left\{\frac{e^{-(x/a)s}}{s^2}\right\}\right]$$

$$= \frac{v_0^2 F_0}{a^2 - v_0^2}\left[\left(t - \frac{x}{v_0}\right)\mathscr{U}\left(t - \frac{x}{v_0}\right) - \left(t - \frac{x}{a}\right)\mathscr{U}\left(t - \frac{x}{a}\right)\right].$$

(b) In the case when $v_0 = a$ the solution of the transformed equation is

$$U(x,s) = c_1 e^{-(s/a)x} + c_2 e^{(s/a)x} - \frac{F_0}{2as}xe^{-(s/a)x}.$$

The usual analysis then leads to $c_1 = 0$ and $c_2 = 0$. Therefore

$$U(x,s) = -\frac{F_0}{2as}xe^{-(s/a)x}$$

and

$$u(x,t) = -\frac{xF_0}{2a}\mathscr{L}^{-1}\left\{\frac{e^{-(x/a)s}}{s}\right\} = -\frac{xF_0}{2a}\mathscr{U}\left(t - \frac{x}{a}\right).$$

──────── **Exercises 14.3** ────────

3. From formulas (5) and (6) in the text,

$$A(\alpha) = \int_0^3 x\cos\alpha x\,dx = \frac{x\sin\alpha x}{\alpha}\Big|_0^3 - \frac{1}{\alpha}\int_0^3 \sin\alpha x\,dx$$

$$= \frac{3\sin 3\alpha}{\alpha} + \frac{\cos\alpha x}{\alpha^2}\Big|_0^3 = \frac{3\alpha\sin 3\alpha + \cos 3\alpha - 1}{\alpha^2}$$

and

$$B(\alpha) = \int_0^3 x\sin\alpha x\,dx = -\frac{x\cos\alpha x}{\alpha}\Big|_0^3 + \frac{1}{\alpha}\int_0^3 \cos\alpha x\,dx$$

$$= -\frac{3\cos 3\alpha}{\alpha} + \frac{\sin\alpha x}{\alpha^2}\Big|_0^3 = \frac{\sin 3\alpha - 3\alpha\cos 3\alpha}{\alpha^2}.$$

Hence

$$f(x) = \frac{1}{\pi}\int_0^\infty \frac{(3\alpha\sin 3\alpha + \cos 3\alpha - 1)\cos\alpha x + (\sin 3\alpha - 3\alpha\cos 3\alpha)\sin\alpha x}{\alpha^2}\,d\alpha$$

$$= \frac{1}{\pi}\int_0^\infty \frac{3\alpha(\sin 3\alpha\cos\alpha x - \cos 3\alpha\sin\alpha x) + \cos 3\alpha\cos\alpha x + \sin 3\alpha\sin\alpha x - \cos\alpha x}{\alpha^2}\,d\alpha$$

$$= \frac{1}{\pi}\int_0^\infty \frac{3\alpha\sin\alpha(3 - x) + \cos\alpha(3 - x) - \cos\alpha x}{\alpha^2}\,d\alpha.$$

193

6. From formulas (5) and (6) in the text,

$$A(\alpha) = \int_{-1}^{1} e^x \cos \alpha x \, dx$$

$$= \frac{e(\cos \alpha + \alpha \sin \alpha) - e^{-1}(\cos \alpha - \alpha \sin \alpha)}{1 + \alpha^2}$$

$$= \frac{2(\sinh 1) \cos \alpha - 2\alpha(\cosh 1) \sin \alpha}{1 + \alpha^2}$$

and

$$B(\alpha) = \int_{-1}^{1} e^x \sin \alpha x \, dx$$

$$= \frac{e(\sin \alpha - \alpha \cos \alpha) - e^{-1}(-\sin \alpha - \alpha \cos \alpha)}{1 + \alpha^2}$$

$$= \frac{2(\cosh 1) \sin \alpha - 2\alpha(\sinh 1) \cos \alpha}{1 + \alpha^2}.$$

Hence

$$f(x) = \frac{1}{\pi} \int_0^\infty [A(\alpha) \cos \alpha x + B(\alpha) \sin \alpha x] \, d\alpha.$$

9. The function is even. Thus from formula (9) in the text

$$A(\alpha) = \int_0^\pi x \cos \alpha x \, dx = \frac{x \sin \alpha x}{\alpha} \bigg|_0^\pi - \frac{1}{\alpha} \int_0^\pi \sin \alpha x \, dx$$

$$= \frac{\pi \alpha \sin \pi \alpha}{\alpha} + \frac{1}{\alpha^2} \cos \alpha x \bigg|_0^\pi = \frac{\pi \alpha \sin \pi \alpha + \cos \pi \alpha - 1}{\alpha^2}.$$

Hence from formula (8) in the text

$$f(x) = \frac{2}{\pi} \int_0^\infty \frac{(\pi \alpha \sin \pi \alpha + \cos \pi \alpha - 1) \cos \alpha x}{\alpha^2} \, d\alpha.$$

12. The function is odd. Thus from formula (11) in the text

$$B(\alpha) = \int_0^\infty x e^{-x} \sin \alpha x \, dx.$$

Now recall

$$\mathscr{L}\{t \sin kt\} = -\frac{d}{ds} \mathscr{L}\{\sin kt\} = 2ks/(s^2 + k^2)^2.$$

If we set $s = 1$ and $k = \alpha$ we obtain

$$B(\alpha) = \frac{2\alpha}{(1 + \alpha^2)^2}.$$

Hence from formula (10) in the text

$$f(x) = \frac{4}{\pi} \int_0^\infty \frac{\alpha \sin \alpha x}{(1 + \alpha^2)^2} \, d\alpha.$$

15. For the cosine integral,

$$A(\alpha) = \int_0^\infty x e^{-2x} \cos \alpha x \, dx.$$

But we know

$$\mathscr{L}\{t \cos kt\} = -\frac{d}{ds} \frac{s}{(s^2 + k^2)} = \frac{(s^2 - k^2)}{(s^2 + k^2)^2}.$$

If we set $s = 2$ and $k = \alpha$ we obtain

$$A(\alpha) = \frac{4 - \alpha^2}{(4 + \alpha^2)^2}.$$

Hence

$$f(x) = \frac{2}{\pi} \int_0^\infty \frac{(4 - \alpha^2) \cos \alpha x}{(4 + \alpha^2)^2} \, d\alpha.$$

For the sine integral,

$$B(\alpha) = \int_0^\infty x e^{-2x} \sin \alpha x \, dx.$$

From Problem 12, we know

$$\mathscr{L}\{t \sin kt\} = \frac{2ks}{(s^2 + k^2)^2}.$$

If we set $s = 2$ and $k = \alpha$ we obtain

$$B(\alpha) = \frac{4\alpha}{(4 + \alpha^2)^2}.$$

Hence

$$f(x) = \frac{8}{\pi} \int_0^\infty \frac{\alpha \sin \alpha x}{(4 + \alpha^2)^2} \, d\alpha.$$

18. From the formula for sine integral of $f(x)$ we have

$$f(x) = \frac{2}{\pi} \int_0^\infty \left(\int_0^\infty f(x) \sin \alpha x \, dx \right) \sin \alpha x \, dx$$

$$= \frac{2}{\pi} \left[\int_0^1 1 \cdot \sin \alpha x \, d\alpha + \int_1^\infty 0 \cdot \sin \alpha x \, d\alpha \right]$$

$$= \frac{2}{\pi} \frac{(-\cos \alpha x)}{x} \Big|_0^1 = \frac{2}{\pi} \frac{1 - \cos x}{x}.$$

Exercises 14.4

For the boundary-value problems in this section it is sometimes useful to note that the identities

$$e^{i\alpha} = \cos \alpha + i \sin \alpha \quad \text{and} \quad e^{-i\alpha} = \cos \alpha - i \sin \alpha$$

imply

$$e^{i\alpha} + e^{-i\alpha} = 2 \cos \alpha \quad \text{and} \quad e^{i\alpha} - e^{-i\alpha} = 2i \sin \alpha.$$

Exercises 14.4

3. Using the Fourier transform, the partial differential equation equation becomes

$$\frac{dU}{dt} + k\alpha^2 U = 0 \qquad \text{and so} \qquad U(\alpha, t) = ce^{-k\alpha^2 t}.$$

Now

$$\mathscr{F}\{u(x,0)\} = U(\alpha, 0) = \sqrt{\pi}\, e^{-\alpha^2/4}$$

by the given result. This gives $c = \sqrt{\pi}\, e^{-\alpha^2/4}$ and so

$$U(\alpha, t) = \sqrt{\pi}\, e^{-(\frac{1}{4}+kt)\alpha^2}.$$

Using the given Fourier transform again we obtain

$$u(x,t) = \sqrt{\pi}\, \mathscr{F}^{-1}\{e^{-(1+4kt)\alpha^2/4}\} = \frac{1}{\sqrt{1+4kt}}\, e^{-x^2/(1+4kt)}.$$

6. The solution of Problem 5 is

$$u(x,t) = \frac{2u_0}{\pi}\int_0^\infty \frac{1-e^{-k\alpha^2 t}}{\alpha}\sin\alpha x\, d\alpha = \frac{2u_0}{\pi}\int_0^\infty \frac{\sin\alpha x}{\alpha}\, d\alpha - \frac{2u_0}{\pi}\int_0^\infty \frac{\sin\alpha x}{\alpha}e^{-k\alpha^2 t}\, d\alpha.$$

Using $\int_0^\infty \frac{\sin\alpha x}{\alpha}\, d\alpha = \pi/2$ this becomes

$$u(x,t) = u_0 - \frac{2u_0}{\pi}\int_0^\infty \frac{\sin\alpha x}{\alpha}e^{-k\alpha^2 t}\, d\alpha.$$

9. Using the Fourier cosine transform we find

$$U(\alpha, t) = ce^{-k\alpha^2 t}.$$

Now

$$\mathscr{F}_C\{u(x,0)\} = \int_0^1 \cos\alpha x\, dx = \frac{\sin\alpha}{\alpha} = U(\alpha, 0).$$

From this we obtain $c = (\sin\alpha)/\alpha$ and so

$$U(\alpha, t) = \frac{\sin\alpha}{\alpha} e^{-k\alpha^2 t}$$

and

$$u(x,t) = \frac{2}{\pi}\int_0^\infty \frac{\sin\alpha}{\alpha} e^{-k\alpha^2 t}\cos\alpha x\, d\alpha.$$

12. Using the Fourier sine transform we obtain

$$U(\alpha, t) = c_1\cos\alpha at + c_2\sin\alpha at.$$

Now

$$\mathscr{F}_S\{u(x,0)\} = \mathscr{F}\{xe^{-x}\} = \int_0^\infty xe^{-x}\sin\alpha x\, dx = \frac{2\alpha}{(1+\alpha^2)^2} = U(\alpha, 0).$$

Also,

$$\mathscr{F}_S\{u_t(x,0)\} = \frac{dU}{dt}\Big|_{t=0} = 0.$$

196

This last condition gives $c_2 = 0$. Then $U(\alpha, 0) = 2\alpha/(1+\alpha^2)^2$ yields $c_1 = 2\alpha/(1+\alpha^2)^2$. Therefore

$$U(\alpha, t) = \frac{2\alpha}{(1+\alpha^2)^2} \cos \alpha a t$$

and

$$u(x, t) = \frac{4}{\pi} \int_0^\infty \frac{\alpha \cos \alpha a t}{(1+\alpha^2)^2} \sin \alpha x \, d\alpha.$$

15. Using the Fourier cosine transform with respect to x gives

$$U(\alpha, y) = c_1 e^{-\alpha y} + c_2 e^{\alpha y}.$$

Since we expect $u(x, y)$ to be bounded as $y \to \infty$ we define $c_2 = 0$. Thus

$$U(\alpha, y) = c_1 e^{-\alpha y}.$$

Now

$$\mathscr{F}_C\{u(x, 0)\} = \int_0^1 50 \cos \alpha x \, dx = 50 \frac{\sin \alpha}{\alpha}$$

and so

$$U(\alpha, y) = 50 \frac{\sin \alpha}{\alpha} e^{-\alpha y}$$

and

$$u(x, y) = \frac{100}{\pi} \int_0^\infty \frac{\sin \alpha}{\alpha} e^{-\alpha y} \cos \alpha x \, d\alpha.$$

18. The domain of y and the boundary condition at $y = 0$ suggest that we use a Fourier cosine transform. The transformed equation is

$$\frac{d^2 U}{dx^2} - \alpha^2 U - u_y(x, 0) = 0 \quad \text{or} \quad \frac{d^2 U}{dx^2} - \alpha^2 U = 0.$$

Because the domain of the variable x is a finite interval we choose to write the general solution of the latter equation as

$$U(x, \alpha) = c_1 \cosh \alpha x + c_2 \sinh \alpha x.$$

Now $U(0, \alpha) = F(\alpha)$, where $F(\alpha)$ is the Fourier cosine transform of $f(y)$, and $U'(\pi, \alpha) = 0$ imply $c_1 = F(\alpha)$ and $c_2 = -F(\alpha) \sinh \alpha \pi / \cosh \alpha \pi$. Thus

$$U(x, \alpha) = F(\alpha) \cosh \alpha x - F(\alpha) \frac{\sinh \alpha \pi}{\cosh \alpha \pi} \sinh \alpha x = F(\alpha) \frac{\cosh \alpha (\pi - x)}{\cosh \alpha \pi}.$$

Using the inverse transform we find that a solution to the problem is

$$u(x, y) = \frac{2}{\pi} \int_0^\infty F(\alpha) \frac{\cosh \alpha (\pi - x)}{\cosh \alpha \pi} \cos \alpha y \, d\alpha.$$

21. Using the Fourier transform with respect to x gives

$$U(\alpha, y) = c_1 \cosh \alpha y + c_2 \sinh \alpha y.$$

The transform of the boundary condition $\dfrac{\partial u}{\partial y}\Big|_{y=0} = 0$ is $\dfrac{dU}{dy}\Big|_{y=0} = 0$. This condition gives $c_2 = 0$. Hence

$$U(\alpha, y) = c_1 \cosh \alpha y.$$

Now by the given information the transform of the boundary condition $u(x, 1) = e^{-x^2}$ is $U(\alpha, 1) = \sqrt{\pi}\, e^{-\alpha^2/4}$. This condition then gives $c_1 = \sqrt{\pi}\, e^{-\alpha^2/4} \cosh \alpha$. Therefore

$$U(\alpha, y) = \sqrt{\pi}\, \frac{e^{-\alpha^2/4} \cosh \alpha y}{\cosh \alpha}$$

and

$$U(x, y) = \frac{1}{2\sqrt{\pi}} \int_{-\infty}^{\infty} \frac{e^{-\alpha^2/4} \cosh \alpha y}{\cosh \alpha}\, e^{-i\alpha x}\, d\alpha$$

$$= \frac{1}{2\sqrt{\pi}} \int_{-\infty}^{\infty} \frac{e^{-\alpha^2/4} \cosh \alpha y}{\cosh \alpha}\, \cos \alpha x\, d\alpha$$

$$= \frac{1}{\sqrt{\pi}} \int_{0}^{\infty} \frac{e^{-\alpha^2/4} \cosh \alpha y}{\cosh \alpha}\, \cos \alpha x\, d\alpha.$$

———— Chapter 14 Review Exercises ————

3. The Laplace transform gives

$$U(x, s) = c_1 e^{-\sqrt{s+h}\, x} + c_2 e^{\sqrt{s+h}\, x} + \frac{u_0}{s+h}.$$

The condition $\lim_{x \to \infty} \partial u/\partial x = 0$ implies $\lim_{x \to \infty} dU/dx = 0$ and so we define $c_2 = 0$. Thus

$$U(x, s) = c_1 e^{-\sqrt{s+h}\, x} + \frac{u_0}{s+h}.$$

The condition $U(0, s) = 0$ then gives $c_1 = -u_0/(s+h)$ and so

$$U(x, s) = \frac{u_0}{s+h} - u_0\, \frac{e^{-\sqrt{s+h}\, x}}{s+h}.$$

With the help of the first translation theorem we then obtain

$$u(x, t) = u_0 \mathscr{L}^{-1}\left\{ \frac{1}{s+h} \right\} - u_0 \mathscr{L}^{-1}\left\{ \frac{e^{-\sqrt{s+h}\, x}}{s+h} \right\} = u_0 e^{-ht} - u_0 e^{-ht} \operatorname{erfc}\left(\frac{x}{2\sqrt{t}} \right)$$

$$= u_0 e^{-ht}\left[1 - \operatorname{erfc}\left(\frac{x}{2\sqrt{t}} \right) \right] = u_0 e^{-ht} \operatorname{erf}\left(\frac{x}{2\sqrt{t}} \right).$$

6. The Laplace transform and undetermined coefficients gives

$$U(x, s) = c_1 \cosh sx + c_2 \sinh sx + \frac{s-1}{s^2 + \pi^2} \sin \pi x.$$

The conditions $U(0, s) = 0$ and $U(1, s) = 0$ give, in turn, $c_1 = 0$ and $c_2 = 0$. Thus

$$U(x, s) = \frac{s - 1}{s^2 + \pi^2} \sin \pi x$$

and

$$u(x, t) = \sin \pi x \, \mathcal{L}^{-1} \left\{ \frac{s}{s^2 + \pi^2} \right\} - \frac{1}{\pi} \sin \pi x \, \mathcal{L}^{-1} \left\{ \frac{\pi}{s^2 + \pi^2} \right\}$$

$$= (\sin \pi x) \cos \pi t - \frac{1}{\pi} (\sin \pi x) \sin \pi t.$$

9. We solve the two problems

$$\frac{\partial^2 u_1}{\partial x^2} + \frac{\partial^2 u_1}{\partial y^2} = 0, \quad x > 0, \quad y > 0,$$

$$u_1(0, y) = 0, \quad y > 0,$$

$$u_1(x, 0) = \begin{cases} 100, & 0 < x < 1 \\ 0, & x > 1 \end{cases}$$

and

$$\frac{\partial^2 u_2}{\partial x^2} + \frac{\partial^2 u_2}{\partial y^2} = 0, \quad x > 0, \quad y > 0,$$

$$u_2(0, y) = \begin{cases} 50, & 0 < y < 1 \\ 0, & y > 1 \end{cases}$$

$$u_2(x, 0) = 0.$$

Using the Fourier sine transform with respect to x we find

$$u_1(x, y) = \frac{200}{\pi} \int_0^\infty \left(\frac{1 - \cos \alpha}{\alpha} \right) e^{-\alpha y} \sin \alpha x \, d\alpha.$$

Using the Fourier sine transform with respect to y we find

$$u_2(x, y) = \frac{100}{\pi} \int_0^\infty \left(\frac{1 - \cos \alpha}{\alpha} \right) e^{-\alpha x} \sin \alpha y \, d\alpha.$$

The solution of the problem is then

$$u(x, y) = u_1(x, y) + u_2(x, y).$$

12. Using the Laplace transform gives

$$U(x, s) = c_1 \cosh \sqrt{s}\, x + c_2 \sinh \sqrt{s}\, x.$$

The condition $u(0, t) = u_0$ transforms into $U(0, s) = u_0/s$. This gives $c_1 = u_0/s$. The condition $u(1, t) = u_0$ transforms into $U(1, s) = u_0/s$. This implies that $c_2 = u_0(1 - \cosh \sqrt{s})/s \sinh \sqrt{s}$.

Hence

$$U(x,s) = \frac{u_0}{s} \cosh \sqrt{s}\, x + u_0 \left[\frac{1 - \cosh \sqrt{s}}{s \sinh \sqrt{s}} \right] \sinh \sqrt{s}\, x$$

$$= u_0 \left[\frac{\sinh \sqrt{s} \cosh \sqrt{s}\, x - \cosh \sinh \sqrt{s} \sinh \sqrt{s}\, x + \sinh \sqrt{s}\, x}{s \sinh \sqrt{s}} \right]$$

$$= u_0 \left[\frac{\sinh \sqrt{s}\, (1 - x) + \sinh \sqrt{s}\, x}{s \sinh \sqrt{s}} \right]$$

$$= u_0 \left[\frac{\sinh \sqrt{s}\, (1 - x)}{s \sinh \sqrt{s}} + \frac{\sinh \sqrt{s}\, x}{s \sinh \sqrt{s}} \right]$$

and

$$u(x,t) = u_0 \left[\mathscr{L}^{-1} \left\{ \frac{\sinh \sqrt{s}\, (1 - x)}{s \sinh \sqrt{s}} \right\} + \mathscr{L}^{-1} \left\{ \frac{\sinh \sqrt{s}\, x}{s \sinh \sqrt{s}} \right\} \right]$$

$$= u_0 \sum_{n=0}^{\infty} \left[\operatorname{erf} \left(\frac{2n + 2 - x}{2\sqrt{t}} \right) - \operatorname{erf} \left(\frac{2n + x}{2\sqrt{t}} \right) \right]$$

$$+ u_0 \sum_{n=0}^{\infty} \left[\operatorname{erf} \left(\frac{2n + 1 + x}{2\sqrt{t}} \right) - \operatorname{erf} \left(\frac{2n + 1 - x}{2\sqrt{t}} \right) \right].$$

15 Numerical Solutions of Partial Differential Equations

_____ **Exercises 15.1** _____

3. The figure shows the values of $u(x, y)$ along the boundary. We need to determine u_{11}, u_{21}, u_{12}, and u_{22}. By symmetry $u_{11} = u_{21}$ and $u_{12} = u_{22}$. The system is

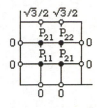

$$u_{21} + u_{12} + 0 + 0 - 4u_{11} = 0$$

$$0 + u_{22} + u_{11} + 0 - 4u_{21} = 0$$

$$u_{22} + \sqrt{3}/2 + 0 + u_{11} - 4u_{12} = 0$$

$$0 + \sqrt{3}/2 + u_{12} + u_{21} - 4u_{22} = 0$$

or

$$3u_{11} + u_{12} = 0$$

$$u_{11} - 3u_{12} = -\frac{\sqrt{3}}{2}.$$

Solving we obtain $u_{11} = u_{21} = \sqrt{3}/16$ and $u_{12} = u_{22} = 3\sqrt{3}/16$.

6. For Gauss-Seidel the coefficients of the unknowns u_{11}, u_{21}, u_{31}, u_{12}, u_{22}, u_{32}, u_{13}, u_{23}, u_{33} are shown in the matrix

$$\begin{bmatrix} 0 & .25 & 0 & .25 & 0 & 0 & 0 & 0 & 0 \\ .25 & 0 & .25 & 0 & .25 & 0 & 0 & 0 & 0 \\ 0 & .25 & 0 & 0 & 0 & .25 & 0 & 0 & 0 \\ .25 & 0 & 0 & 0 & .25 & 0 & .25 & 0 & 0 \\ 0 & .25 & 0 & .25 & 0 & .25 & 0 & .25 & 0 \\ 0 & 0 & .25 & 0 & .25 & 0 & 0 & 0 & .25 \\ 0 & 0 & 0 & .25 & 0 & 0 & 0 & .25 & 0 \\ 0 & 0 & 0 & 0 & .25 & 0 & .25 & 0 & .25 \\ 0 & 0 & 0 & 0 & 0 & .25 & 0 & .25 & 0 \end{bmatrix}.$$

The constant terms are 7.5, 5, 20, 10, 0, 15, 17.5, 5, 27.5. We use 32.5 as the initial guess for each variable. Then $u_{11} = 21.92$, $u_{21} = 28.30$, $u_{31} = 38.17$, $u_{12} = 29.38$, $u_{22} = 33.13$, $u_{32} = 44.38$, $u_{13} = 22.46$, $u_{23} = 30.45$, and $u_{33} = 46.21$.

Exercises 15.2

3. We identify $c = 1$, $a = 2$, $T = 1$, $n = 8$, and $m = 40$. Then $h = 2/8 = 0.25$, $k = 1/40 = 0.025$, and $\lambda = 2/5 = 0.4$.

TIME	X=0.25	X=0.50	X=0.75	X=1.00	X=1.25	X=1.50	X=1.75
0.000	1.0000	1.0000	1.0000	1.0000	0.0000	0.0000	0.0000
0.025	0.7074	0.9520	0.9566	0.7444	0.2545	0.0371	0.0053
0.050	0.5606	0.8499	0.8685	0.6633	0.3303	0.1034	0.0223
0.075	0.4684	0.7473	0.7836	0.6191	0.3614	0.1529	0.0462
0.100	0.4015	0.6577	0.7084	0.5837	0.3753	0.1871	0.0684
0.125	0.3492	0.5821	0.6428	0.5510	0.3797	0.2101	0.0861
0.150	0.3069	0.5187	0.5857	0.5199	0.3778	0.2247	0.0990
0.175	0.2721	0.4652	0.5359	0.4901	0.3716	0.2329	0.1078
0.200	0.2430	0.4198	0.4921	0.4617	0.3622	0.2362	0.1132
0.225	0.2186	0.3809	0.4533	0.4348	0.3507	0.2358	0.1160
0.250	0.1977	0.3473	0.4189	0.4093	0.3378	0.2327	0.1166
0.275	0.1798	0.3181	0.3881	0.3853	0.3240	0.2275	0.1157
0.300	0.1643	0.2924	0.3604	0.3626	0.3097	0.2208	0.1136
0.325	0.1507	0.2697	0.3353	0.3412	0.2953	0.2131	0.1107
0.350	0.1387	0.2495	0.3125	0.3211	0.2808	0.2047	0.1071
0.375	0.1281	0.2313	0.2916	0.3021	0.2666	0.1960	0.1032
0.400	0.1187	0.2150	0.2725	0.2843	0.2528	0.1871	0.0989
0.425	0.1102	0.2002	0.2549	0.2675	0.2393	0.1781	0.0946
0.450	0.1025	0.1867	0.2387	0.2517	0.2263	0.1692	0.0902
0.475	0.0955	0.1743	0.2236	0.2368	0.2139	0.1606	0.0858
0.500	0.0891	0.1630	0.2097	0.2228	0.2020	0.1521	0.0814
0.525	0.0833	0.1525	0.1967	0.2096	0.1906	0.1439	0.0772
0.550	0.0779	0.1429	0.1846	0.1973	0.1798	0.1361	0.0731
0.575	0.0729	0.1339	0.1734	0.1856	0.1696	0.1285	0.0691
0.600	0.0683	0.1256	0.1628	0.1746	0.1598	0.1214	0.0653
0.625	0.0641	0.1179	0.1530	0.1643	0.1506	0.1145	0.0617
0.650	0.0601	0.1106	0.1438	0.1546	0.1419	0.1080	0.0582
0.675	0.0564	0.1039	0.1351	0.1455	0.1336	0.1018	0.0549
0.700	0.0530	0.0976	0.1270	0.1369	0.1259	0.0959	0.0518
0.725	0.0497	0.0917	0.1194	0.1288	0.1185	0.0904	0.0488
0.750	0.0467	0.0862	0.1123	0.1212	0.1116	0.0852	0.0460
0.775	0.0439	0.0810	0.1056	0.1140	0.1050	0.0802	0.0433
0.800	0.0413	0.0762	0.0993	0.1073	0.0989	0.0755	0.0408
0.825	0.0388	0.0716	0.0934	0.1009	0.0931	0.0711	0.0384
0.850	0.0365	0.0674	0.0879	0.0950	0.0876	0.0669	0.0362
0.875	0.0343	0.0633	0.0827	0.0894	0.0824	0.0630	0.0341
0.900	0.0323	0.0596	0.0778	0.0841	0.0776	0.0593	0.0321
0.925	0.0303	0.0560	0.0732	0.0791	0.0730	0.0558	0.0302
0.950	0.0285	0.0527	0.0688	0.0744	0.0687	0.0526	0.0284
0.975	0.0268	0.0496	0.0647	0.0700	0.0647	0.0495	0.0268
1.000	0.0253	0.0466	0.0609	0.0659	0.0608	0.0465	0.0252

(x,y)	exact	approx	abs error
(0.25,0.1)	0.3794	0.4015	0.0221
(1,0.5)	0.1854	0.2228	0.0374
(1.5,0.8)	0.0623	0.0755	0.0132

6. (a) We identify $c = 15/88 \approx 0.1705$, $a = 20$, $T = 10$, $n = 10$, and $m = 10$. Then $h = 2$, $k = 1$, and $\lambda = 15/352 \approx 0.0426$.

TIME	X=2	X=4	X=6	X=8	X=10	X=12	X=14	X=16	X=18
0	30.0000	30.0000	30.0000	30.0000	30.0000	30.0000	30.0000	30.0000	30.0000
1	28.7216	30.0000	30.0000	30.0000	30.0000	30.0000	30.0000	30.0000	28.7216
2	27.5521	29.9455	30.0000	30.0000	30.0000	30.0000	30.0000	29.9455	27.5521
3	26.4800	29.8459	29.9977	30.0000	30.0000	30.0000	29.9977	29.8459	26.4800
4	25.4951	29.7089	29.9913	29.9999	30.0000	29.9999	29.9913	29.7089	25.4951
5	24.5882	29.5414	29.9796	29.9995	30.0000	29.9995	29.9796	29.5414	24.5882
6	23.7515	29.3490	29.9618	29.9987	30.0000	29.9987	29.9618	29.3490	23.7515
7	22.9779	29.1365	29.9373	29.9972	29.9998	29.9972	29.9373	29.1365	22.9779
8	22.2611	28.9082	29.9057	29.9948	29.9996	29.9948	29.9057	28.9082	22.2611
9	21.5958	28.6675	29.8670	29.9912	29.9992	29.9912	29.8670	28.6675	21.5958
10	20.9768	28.4172	29.8212	29.9862	29.9985	29.9862	29.8212	28.4172	20.9768

(b) We identify $c = 15/88 \approx 0.1705$, $a = 50$, $T = 10$, $n = 10$, and $m = 10$. Then $h = 5$, $k = 1$, and $\lambda = 3/440 \approx 0.0068$.

TIME	X=5	X=10	X=15	X=20	X=25	X=30	X=35	X=40	X=45
0	30.0000	30.0000	30.0000	30.0000	30.0000	30.0000	30.0000	30.0000	30.0000
1	29.7955	30.0000	30.0000	30.0000	30.0000	30.0000	30.0000	30.0000	29.7955
2	29.5937	29.9986	30.0000	30.0000	30.0000	30.0000	30.0000	29.9986	29.5937
3	29.3947	29.9959	30.0000	30.0000	30.0000	30.0000	30.0000	29.9959	29.3947
4	29.1984	29.9918	30.0000	30.0000	30.0000	30.0000	30.0000	29.9918	29.1984
5	29.0047	29.9864	29.9999	30.0000	30.0000	30.0000	29.9999	29.9864	29.0047
6	28.8136	29.9798	29.9998	30.0000	30.0000	30.0000	29.9998	29.9798	28.8136
7	28.6251	29.9720	29.9997	30.0000	30.0000	30.0000	29.9997	29.9720	28.6251
8	28.4391	29.9630	29.9995	30.0000	30.0000	30.0000	29.9995	29.9630	28.4391
9	28.2556	29.9529	29.9992	30.0000	30.0000	30.0000	29.9992	29.9529	28.2556
10	28.0745	29.9416	29.9989	30.0000	30.0000	30.0000	29.9989	29.9416	28.0745

(c) We identify $c = 50/27 \approx 1.8519$, $a = 20$, $T = 10$, $n = 10$, and $m = 10$. Then $h = 2$, $k = 1$, and $\lambda = 25/54 \approx 0.4630$.

TIME	X=2	X=4	X=6	X=8	X=10	X=12	X=14	X=16	X=18
0	18.0000	32.0000	42.0000	48.0000	50.0000	48.0000	42.0000	32.0000	18.0000
1	16.1481	30.1481	40.1481	46.1481	48.1481	46.1481	40.1481	30.1481	16.1481
2	15.1536	28.2963	38.2963	44.2963	46.2963	44.2963	38.2963	28.2963	15.1536
3	14.2226	26.8414	36.4444	42.4444	44.4444	42.4444	36.4444	26.8414	14.2226
4	13.4801	25.4452	34.7764	40.5926	42.5926	40.5926	34.7764	25.4452	13.4801
5	12.7787	24.2258	33.1491	38.8258	40.7407	38.8258	33.1491	24.2258	12.7787
6	12.1622	23.0574	31.6460	37.0842	38.9677	37.0842	31.6460	23.0574	12.1622
7	11.5756	21.9895	30.1875	35.4385	37.2238	35.4385	30.1875	21.9895	11.5756
8	11.0378	20.9636	28.8232	33.8340	35.5707	33.8340	28.8232	20.9636	11.0378
9	10.5230	20.0070	27.5043	32.3182	33.9626	32.3182	27.5043	20.0070	10.5230
10	10.0420	19.0872	26.2620	30.8509	32.4400	30.8509	26.2620	19.0872	10.0420

(d) We identify $c = 260/159 \approx 1.6352$, $a = 100$, $T = 10$, $n = 10$, and $m = 10$. Then $h = 10$, $k = 1$, and $\lambda = 13/795 \approx 00164$.

TIME	X=10	X=20	X=30	X=40	X=50	X=60	X=70	X=80	X=90
0	8.0000	16.0000	24.0000	32.0000	40.0000	32.0000	24.0000	16.0000	8.0000
1	8.0000	16.0000	23.6075	31.3459	39.2151	31.6075	23.7384	15.8692	8.0000
2	8.0000	15.9936	23.2279	30.7068	38.4452	31.2151	23.4789	15.7384	7.9979
3	7.9999	15.9812	22.8606	30.0824	37.6900	30.8229	23.2214	15.6076	7.9937
4	7.9996	15.9631	22.5050	29.4724	36.9492	30.4312	22.9660	15.4769	7.9874
5	7.9990	15.9399	22.1606	28.8765	36.2228	30.0401	22.7125	15.3463	7.9793
6	7.9981	15.9118	21.8270	28.2945	35.5103	29.6500	22.4610	15.2158	7.9693
7	7.9967	15.8791	21.5037	27.7261	34.8117	29.2610	22.2112	15.0854	7.9575
8	7.9948	15.8422	21.1902	27.1709	34.1266	28.8733	21.9633	14.9553	7.9439
9	7.9924	15.8013	20.8861	26.6288	33.4548	28.4870	21.7172	14.8253	7.9287
10	7.9894	15.7568	20.5911	26.0995	32.7961	28.1024	21.4727	14.6956	7.9118

9. (a) We identify $c = 15/88 \approx 0.1705$, $a = 20$, $T = 10$, $n = 10$, and $m = 10$. Then $h = 2$, $k = 1$, and $\lambda = 15/352 \approx 0.0426$.

TIME	X=2.00	X=4.00	X=6.00	X=8.00	X=10.00	X=12.00	X=14.00	X=16.00	X=18.00
0.00	30.0000	30.0000	30.0000	30.0000	30.0000	30.0000	30.0000	30.0000	30.0000
1.00	28.7733	29.9749	29.9995	30.0000	30.0000	30.0000	29.9998	29.9916	29.5911
2.00	27.6450	29.9037	29.9970	29.9999	30.0000	30.0000	29.9990	29.9679	29.2150
3.00	26.6051	29.7938	29.9911	29.9997	30.0000	29.9999	29.9970	29.9313	28.8684
4.00	25.6452	29.6517	29.9805	29.9991	30.0000	29.9997	29.9935	29.8839	28.5484
5.00	24.7573	29.4829	29.9643	29.9981	29.9999	29.9994	29.9881	29.8276	28.2524
6.00	23.9347	29.2922	29.9421	29.9963	29.9997	29.9988	29.9807	29.7641	27.9782
7.00	23.1711	29.0836	29.9134	29.9936	29.9995	29.9979	29.9711	29.6945	27.7237
8.00	22.4612	28.8606	29.8782	29.9899	29.9991	29.9966	29.9594	29.6202	27.4870
9.00	21.7999	28.6263	29.8362	29.9848	29.9985	29.9949	29.9454	29.5421	27.2666
10.00	21.1829	28.3831	29.7878	29.9783	29.9976	29.9927	29.9293	29.4610	27.0610

(b) We identify $c = 15/88 \approx 0.1705$, $a = 50$, $T = 10$, $n = 10$, and $m = 10$. Then $h = 5$, $k = 1$, and $\lambda = 3/440 \approx 0.0068$.

TIME	X=5.00	X=10.00	X=15.00	X=20.00	X=25.00	X=30.00	X=35.00	X=40.00	X=45.00
0.00	30.0000	30.0000	30.0000	30.0000	30.0000	30.0000	30.0000	30.0000	30.0000
1.00	29.7968	29.9993	30.0000	30.0000	30.0000	30.0000	30.0000	29.9998	29.9323
2.00	29.5964	29.9973	30.0000	30.0000	30.0000	30.0000	30.0000	29.9991	29.8655
3.00	29.3987	29.9939	30.0000	30.0000	30.0000	30.0000	30.0000	29.9980	29.7996
4.00	29.2036	29.9893	29.9999	30.0000	30.0000	30.0000	30.0000	29.9964	29.7345
5.00	29.0112	29.9834	29.9998	30.0000	30.0000	30.0000	29.9999	29.9945	29.6704
6.00	28.8212	29.9762	29.9997	30.0000	30.0000	30.0000	29.9999	29.9921	29.6071
7.00	28.6339	29.9679	29.9995	30.0000	30.0000	30.0000	29.9998	29.9893	29.5446
8.00	28.4490	29.9585	29.9992	30.0000	30.0000	30.0000	29.9997	29.9862	29.4830
9.00	28.2665	29.9479	29.9989	30.0000	30.0000	30.0000	29.9996	29.9827	29.4222
10.00	28.0864	29.9363	29.9986	30.0000	30.0000	30.0000	29.9995	29.9788	29.3621

(c) We identify $c = 50/27 \approx 1.8519$, $a = 20$, $T = 10$, $n = 10$, and $m = 10$. Then $h = 2$, $k = 1$, and $\lambda = 25/54 \approx 0.4630$.

TIME	X=2.00	X=4.00	X=6.00	X=8.00	X=10.00	X=12.00	X=14.00	X=16.00	X=18.00
0.00	18.0000	32.0000	42.0000	48.0000	50.0000	48.0000	42.0000	32.0000	18.0000
1.00	16.4489	30.1970	40.1562	46.1502	48.1531	46.1773	40.3274	31.2520	22.9449
2.00	15.3312	28.5350	38.3477	44.3130	46.3327	44.4671	39.0872	31.5755	24.6930
3.00	14.4219	27.0429	36.6090	42.5113	44.5759	42.9362	38.1976	31.7478	25.4131
4.00	13.6381	25.6913	34.9606	40.7728	42.9127	41.5716	37.4340	31.7086	25.6986
5.00	12.9409	24.4545	33.4091	39.1182	41.3519	40.3240	36.7033	31.5136	25.7663
6.00	12.3088	23.3146	31.9546	37.5566	39.8880	39.1565	35.9745	31.2134	25.7128
7.00	11.7294	22.2589	30.5939	36.0884	38.5109	38.0470	35.2407	30.8434	25.5871
8.00	11.1946	21.2785	29.3217	34.7092	37.2109	36.9834	34.5032	30.4279	25.4167
9.00	10.6987	20.3660	28.1318	33.4130	35.9801	35.9591	33.7660	29.9836	25.2181
10.00	10.2377	19.5150	27.0178	32.1929	34.8117	34.9710	33.0338	29.5224	25.0019

(d) We identify $c = 260/159 \approx 1.6352$, $a = 100$, $T = 10$, $n = 10$, and $m = 10$. Then $h = 10$, $k = 1$, and $\lambda = 13/795 \approx 00164$.

TIME	X=10.00	X=20.00	X=30.00	X=40.00	X=50.00	X=60.00	X=70.00	X=80.00	X=90.00
0.00	8.0000	16.0000	24.0000	32.0000	40.0000	32.0000	24.0000	16.0000	8.0000
1.00	8.0000	16.0000	24.0000	31.9979	39.7425	31.9979	24.0000	16.0026	8.3218
2.00	8.0000	16.0000	23.9999	31.9918	39.4932	31.9918	24.0000	16.0102	8.6333
3.00	8.0000	16.0000	23.9997	31.9820	39.2517	31.9820	24.0001	16.0225	8.9350
4.00	8.0000	16.0000	23.9993	31.9687	39.0176	31.9687	24.0002	16.0392	9.2272
5.00	8.0000	16.0000	23.9987	31.9520	38.7905	31.9521	24.0003	16.0599	9.5103
6.00	8.0000	15.9999	23.9978	31.9323	38.5701	31.9324	24.0005	16.0845	9.7846
7.00	8.0000	15.9999	23.9966	31.9097	38.3561	31.9098	24.0008	16.1126	10.0506
8.00	8.0000	15.9998	23.9951	31.8844	38.1483	31.8846	24.0012	16.1441	10.3084
9.00	8.0000	15.9997	23.9931	31.8566	37.9463	31.8569	24.0017	16.1786	10.5585
10.00	8.0000	15.9996	23.9908	31.8265	37.7499	31.8270	24.0023	16.2160	10.8012

12. We identify $c = 1$, $a = 1$, $T = 1$, $n = 5$, and $m = 20$. Then $h = 0.2$, $k = 0.04$, and $\lambda = 1$. The values below were obtained using *Excel*, which carries more than 12 significant digits. In order to see evidence of instability use $0 \le t \le 2$.

TIME	X=0.2	X=0.4	X=0.6	X=0.8
0.00	0.5878	0.9511	0.9511	0.5878
0.04	0.3633	0.5878	0.5878	0.3633
0.08	0.2245	0.3633	0.3633	0.2245
0.12	0.1388	0.2245	0.2245	0.1388
0.16	0.0858	0.1388	0.1388	0.0858
0.20	0.0530	0.0858	0.0858	0.0530
0.24	0.0328	0.0530	0.0530	0.0328
0.28	0.0202	0.0328	0.0328	0.0202
0.32	0.0125	0.0202	0.0202	0.0125
0.36	0.0077	0.0125	0.0125	0.0077
0.40	0.0048	0.0077	0.0077	0.0048
0.44	0.0030	0.0048	0.0048	0.0030
0.48	0.0018	0.0030	0.0030	0.0018

(continued)

205

Exercises 15.2

(continued)

0.52	0.0011	0.0018	0.0018	0.0011
0.56	0.0007	0.0011	0.0011	0.0007
0.60	0.0004	0.0007	0.0007	0.0004
0.64	0.0003	0.0004	0.0004	0.0003
0.68	0.0002	0.0003	0.0003	0.0002
0.72	0.0001	0.0002	0.0002	0.0001
0.76	0.0001	0.0001	0.0001	0.0001
0.80	0.0000	0.0001	0.0001	0.0000
0.84	0.0000	0.0000	0.0000	0.0000
0.88	0.0000	0.0000	0.0000	0.0000
0.92	0.0000	0.0000	0.0000	0.0000
0.96	0.0000	0.0000	0.0000	0.0000
1.00	0.0000	0.0000	0.0000	0.0000
1.04	0.0000	0.0000	0.0000	0.0000
1.08	0.0000	0.0000	0.0000	0.0000
1.12	0.0000	0.0000	0.0000	0.0000
1.16	0.0000	0.0000	0.0000	0.0000
1.20	−0.0001	0.0001	−0.0001	0.0001
1.24	0.0001	−0.0002	0.0002	−0.0001
1.28	−0.0004	0.0006	−0.0006	0.0004
1.32	0.0010	−0.0015	0.0015	−0.0010
1.36	−0.0025	0.0040	−0.0040	0.0025
1.40	0.0065	−0.0106	0.0106	−0.0065
1.44	−0.0171	0.0277	−0.0277	0.0171
1.48	0.0448	−0.0724	0.0724	−0.0448
1.52	−0.1172	0.1897	−0.1897	0.1172
1.56	0.3069	−0.4965	0.4965	−0.3069
1.60	−0.8034	1.2999	−1.2999	0.8034
1.64	2.1033	−3.4032	3.4032	−2.1033
1.68	−5.5064	8.9096	−8.9096	5.5064
1.72	14.416	−23.326	23.326	−14.416
1.76	−37.742	61.067	−61.067	37.742
1.80	98.809	−159.88	159.88	−98.809
1.84	−258.68	418.56	−418.56	258.685
1.88	677.24	−1095.8	1095.8	−677.245
1.92	−1773.1	2868.9	−2868.9	1773.1
1.96	4641.9	−7510.8	7510.8	−4641.9
2.00	−12153	19663	−19663	12153

_____ **Exercises 15.3** _____

3. (a) Identifying $h = 1/5$ and $k = 0.5/10 = 0.05$ we see that $\lambda = 0.25$.

TIME	X=0.2	X=0.4	X=0.6	X=0.8
0.00	0.5878	0.9511	0.9511	0.5878
0.05	0.5808	0.9397	0.9397	0.5808
0.10	0.5599	0.9059	0.9059	0.5599
0.15	0.5256	0.8505	0.8505	0.5256
0.20	0.4788	0.7748	0.7748	0.4788
0.25	0.4206	0.6806	0.6806	0.4206
0.30	0.3524	0.5701	0.5701	0.3524
0.35	0.2757	0.4460	0.4460	0.2757
0.40	0.1924	0.3113	0.3113	0.1924
0.45	0.1046	0.1692	0.1692	0.1046
0.50	0.0142	0.0230	0.0230	0.0142

(b) Identifying $h = 1/5$ and $k = 0.5/20 = 0.025$ we see that $\lambda = 0.125$.

TIME	X=0.2	X=0.4	X=0.6	X=0.8
0.00	0.5878	0.9511	0.9511	0.5878
0.03	0.5860	0.9482	0.9482	0.5860
0.05	0.5808	0.9397	0.9397	0.5808
0.08	0.5721	0.9256	0.9256	0.5721
0.10	0.5599	0.9060	0.9060	0.5599
0.13	0.5445	0.8809	0.8809	0.5445
0.15	0.5257	0.8507	0.8507	0.5257
0.18	0.5039	0.8153	0.8153	0.5039
0.20	0.4790	0.7750	0.7750	0.4790
0.23	0.4513	0.7302	0.7302	0.4513
0.25	0.4209	0.6810	0.6810	0.4209
0.28	0.3879	0.6277	0.6277	0.3879
0.30	0.3527	0.5706	0.5706	0.3527
0.33	0.3153	0.5102	0.5102	0.3153
0.35	0.2761	0.4467	0.4467	0.2761
0.38	0.2352	0.3806	0.3806	0.2352
0.40	0.1929	0.3122	0.3122	0.1929
0.43	0.1495	0.2419	0.2419	0.1495
0.45	0.1052	0.1701	0.1701	0.1052
0.48	0.0602	0.0974	0.0974	0.0602
0.50	0.0149	0.0241	0.0241	0.0149

6. We identify $c = 24944.4$, $k = 0.00010022$ seconds $= 0.10022$ milliseconds, and $\lambda = 0.25$. Time in the table is expressed in milliseconds.

TIME	X=10	X=20	X=30	X=40	X=50
0.00000	0.2000	0.2667	0.2000	0.1333	0.0667
0.10022	0.1958	0.2625	0.2000	0.1333	0.0667
0.20045	0.1836	0.2503	0.1997	0.1333	0.0667
0.30067	0.1640	0.2307	0.1985	0.1333	0.0667
0.40089	0.1384	0.2050	0.1952	0.1332	0.0667
0.50111	0.1083	0.1744	0.1886	0.1328	0.0667
0.60134	0.0755	0.1407	0.1777	0.1318	0.0666
0.70156	0.0421	0.1052	0.1615	0.1295	0.0665
0.80178	0.0100	0.0692	0.1399	0.1253	0.0661
0.90201	-0.0190	0.0340	0.1129	0.1184	0.0654
1.00223	-0.0435	0.0004	0.0813	0.1077	0.0638
1.10245	-0.0626	-0.0309	0.0464	0.0927	0.0610
1.20268	-0.0758	-0.0593	0.0095	0.0728	0.0564
1.30290	-0.0832	-0.0845	-0.0278	0.0479	0.0493
1.40312	-0.0855	-0.1060	-0.0639	0.0184	0.0390
1.50334	-0.0837	-0.1237	-0.0974	-0.0150	0.0250
1.60357	-0.0792	-0.1371	-0.1275	-0.0511	0.0069
1.70379	-0.0734	-0.1464	-0.1533	-0.0882	-0.0152
1.80401	-0.0675	-0.1515	-0.1747	-0.1249	-0.0410
1.90424	-0.0627	-0.1528	-0.1915	-0.1595	-0.0694
2.00446	-0.0596	-0.1509	-0.2039	-0.1904	-0.0991
2.10468	-0.0585	-0.1467	-0.2122	-0.2165	-0.1283
2.20491	-0.0592	-0.1410	-0.2166	-0.2368	-0.1551
2.30513	-0.0614	-0.1349	-0.2175	-0.2507	-0.1772
2.40535	-0.0643	-0.1294	-0.2154	-0.2579	-0.1929
2.50557	-0.0672	-0.1251	-0.2105	-0.2585	-0.2005
2.60580	-0.0696	-0.1227	-0.2033	-0.2524	-0.1993
2.70602	-0.0709	-0.1219	-0.1942	-0.2399	-0.1889
2.80624	-0.0710	-0.1225	-0.1833	-0.2214	-0.1699
2.90647	-0.0699	-0.1236	-0.1711	-0.1972	-0.1435
3.00669	-0.0678	-0.1244	-0.1575	-0.1681	-0.1115
3.10691	-0.0649	-0.1237	-0.1425	-0.1348	-0.0761
3.20713	-0.0617	-0.1205	-0.1258	-0.0983	-0.0395
3.30736	-0.0583	-0.1139	-0.1071	-0.0598	-0.0042
3.40758	-0.0547	-0.1035	-0.0859	-0.0209	0.0279
3.50780	-0.0508	-0.0889	-0.0617	0.0171	0.0552
3.60803	-0.0460	-0.0702	-0.0343	0.0525	0.0767
3.70825	-0.0399	-0.0478	-0.0037	0.0840	0.0919
3.80847	-0.0318	-0.0221	0.0297	0.1106	0.1008
3.90870	-0.0211	0.0062	0.0648	0.1314	0.1041
4.00892	-0.0074	0.0365	0.1005	0.1464	0.1025
4.10914	0.0095	0.0680	0.1350	0.1558	0.0973
4.20936	0.0295	0.1000	0.1666	0.1602	0.0897
4.30959	0.0521	0.1318	0.1937	0.1606	0.0808
4.40981	0.0764	0.1625	0.2148	0.1581	0.0719
4.51003	0.1013	0.1911	0.2291	0.1538	0.0639
4.61026	0.1254	0.2164	0.2364	0.1485	0.0575
4.71048	0.1475	0.2373	0.2369	0.1431	0.0532
4.81070	0.1659	0.2526	0.2315	0.1379	0.0512
4.91093	0.1794	0.2611	0.2217	0.1331	0.0514
5.01115	0.1867	0.2620	0.2087	0.1288	0.0535

Chapter 15 Review Exercises

3. (a)

TIME	X=0.0	X=0.2	X=0.4	X=0.6	X=0.8	X=1.0
0.00	0.0000	0.2000	0.4000	0.6000	0.8000	0.0000
0.01	0.0000	0.2000	0.4000	0.6000	0.5500	0.0000
0.02	0.0000	0.2000	0.4000	0.5375	0.4250	0.0000
0.03	0.0000	0.2000	0.3844	0.4750	0.3469	0.0000
0.04	0.0000	0.1961	0.3609	0.4203	0.2922	0.0000
0.05	0.0000	0.1883	0.3346	0.3734	0.2512	0.0000

(b)

TIME	X=0.0	X=0.2	X=0.4	X=0.6	X=0.8	X=1.0
0.00	0.0000	0.2000	0.4000	0.6000	0.8000	0.0000
0.01	0.0000	0.2000	0.4000	0.6000	0.8000	0.0000
0.02	0.0000	0.2000	0.4000	0.6000	0.5500	0.0000
0.03	0.0000	0.2000	0.4000	0.5375	0.4250	0.0000
0.04	0.0000	0.2000	0.3844	0.4750	0.3469	0.0000
0.05	0.0000	0.1961	0.3609	0.4203	0.2922	0.0000

(c) The table in part (b) is the same as the table in part (a) shifted downward one row.

Appendix I

---------- **Gamma Function** ----------

3. If $t = x^3$, then $dt = 3x^2 \, dx$ and $x^4 \, dx = \frac{1}{3} t^{2/3} \, dt$. Now

$$\int_0^\infty x^4 e^{-x^3} \, dx = \int_0^\infty \frac{1}{3} t^{2/3} e^{-t} \, dt = \frac{1}{3} \int_0^\infty t^{2/3} e^{-t} \, dt$$

$$= \frac{1}{3} \Gamma\left(\frac{5}{3}\right) = \frac{1}{3}(0.89) \approx 0.297.$$

6. For $x > 0$

$$\Gamma(x+1) = \int_0^\infty t^x e^{-t} dt$$

$$\begin{array}{ll} u = t^x & dv = e^{-t} \, dt \\ du = xt^{x-1} \, dt & v = -e^{-t} \end{array}$$

$$= -t^x e^{-t} \Big|_0^\infty - \int_0^\infty xt^{x-1}(-e^{-t}) \, dt$$

$$= x \int_0^\infty t^{x-1} e^{-t} dt = x\Gamma(x).$$

Appendix II

---------- **Introduction to Matrices** ----------

3. (a) $\mathbf{AB} = \begin{pmatrix} -2 - 9 & 12 - 6 \\ 5 + 12 & -30 + 8 \end{pmatrix} = \begin{pmatrix} -11 & 6 \\ 17 & -22 \end{pmatrix}$

(b) $\mathbf{BA} = \begin{pmatrix} -2 - 30 & 3 + 24 \\ 6 - 10 & -9 + 8 \end{pmatrix} = \begin{pmatrix} -32 & 27 \\ -4 & -1 \end{pmatrix}$

(c) $\mathbf{A}^2 = \begin{pmatrix} 4 + 15 & -6 - 12 \\ -10 - 20 & 15 + 16 \end{pmatrix} = \begin{pmatrix} 19 & -18 \\ -30 & 31 \end{pmatrix}$

(d) $\mathbf{B}^2 = \begin{pmatrix} 1 + 18 & -6 + 12 \\ -3 + 6 & 18 + 4 \end{pmatrix} = \begin{pmatrix} 19 & 6 \\ 3 & 22 \end{pmatrix}$

6. (a) $\mathbf{AB} = (5 \quad -6 \quad 7) \begin{pmatrix} 3 \\ 4 \\ -1 \end{pmatrix} = (-16)$

(b) $\mathbf{BA} = \begin{pmatrix} 3 \\ 4 \\ -1 \end{pmatrix} (5 \quad -6 \quad 7) = \begin{pmatrix} 15 & -18 & 21 \\ 20 & -24 & 28 \\ -5 & 6 & -7 \end{pmatrix}$

(c) $(\mathbf{BA})\mathbf{C} = \begin{pmatrix} 15 & -18 & 21 \\ 20 & -24 & 28 \\ -5 & 6 & -7 \end{pmatrix} \begin{pmatrix} 1 & 2 & 4 \\ 0 & 1 & -1 \\ 3 & 2 & 1 \end{pmatrix} = \begin{pmatrix} 78 & 54 & 99 \\ 104 & 72 & 132 \\ -26 & -18 & -33 \end{pmatrix}$

(d) Since $\mathbf{AB}$ is 1×1 and $\mathbf{C}$ is 3×3 the product $(\mathbf{AB})\mathbf{C}$ is not defined.

9. (a) $(\mathbf{AB})^T = \begin{pmatrix} 7 & 10 \\ 38 & 75 \end{pmatrix}^T = \begin{pmatrix} 7 & 38 \\ 10 & 75 \end{pmatrix}$

(b) $\mathbf{B}^T\mathbf{A}^T = \begin{pmatrix} 5 & -2 \\ 10 & -5 \end{pmatrix} \begin{pmatrix} 3 & 8 \\ 4 & 1 \end{pmatrix} = \begin{pmatrix} 7 & 38 \\ 10 & 75 \end{pmatrix}$

12. $\begin{pmatrix} 6t \\ 3t^2 \\ -3t \end{pmatrix} + \begin{pmatrix} -t+1 \\ -t^2+t \\ 3t-3 \end{pmatrix} - \begin{pmatrix} 6t \\ 8 \\ -10t \end{pmatrix} = \begin{pmatrix} -t+1 \\ 2t^2+t-8 \\ 10t-3 \end{pmatrix}$

15. Since $\det \mathbf{A} = 0$, $\mathbf{A}$ is singular.

18. Since $\det \mathbf{A} = -6$, $\mathbf{A}$ is nonsingular.

$$\mathbf{A}^{-1} = -\frac{1}{6} \begin{pmatrix} 2 & -10 \\ -2 & 7 \end{pmatrix}$$

21. Since $\det \mathbf{A} = -9$, $\mathbf{A}$ is nonsingular. The cofactors are

$$\begin{array}{lll} A_{11} = -2 & A_{12} = -13 & A_{13} = 8 \\ A_{21} = -2 & A_{22} = 5 & A_{23} = -1 \\ A_{31} = -1 & A_{32} = 7 & A_{33} = -5. \end{array}$$

Then

$$\mathbf{A}^{-1} = -\frac{1}{9} \begin{pmatrix} -2 & -13 & 8 \\ -2 & 5 & -1 \\ -1 & 7 & -5 \end{pmatrix}^T = -\frac{1}{9} \begin{pmatrix} -2 & -2 & -1 \\ -13 & 5 & 7 \\ 8 & -1 & -5 \end{pmatrix}.$$

24. Since $\det \mathbf{A}(t) = 2e^{2t} \neq 0$, $\mathbf{A}$ is nonsingular.

$$\mathbf{A}^{-1} = \frac{1}{2}e^{-2t} \begin{pmatrix} e^t \sin t & 2e^t \cos t \\ -e^t \cos t & 2e^t \sin t \end{pmatrix}$$

27. $\mathbf{X} = \begin{pmatrix} 2e^{2t} + 8e^{-3t} \\ -2e^{2t} + 4e^{-3t} \end{pmatrix}$ so that $\dfrac{d\mathbf{X}}{dt} = \begin{pmatrix} 4e^{2t} - 24e^{-3t} \\ -4e^{2t} - 12e^{-3t} \end{pmatrix}.$

30. (a) $\dfrac{d\mathbf{A}}{dt} = \begin{pmatrix} -2t/(t^2+1)^2 & 3 \\ 2t & 1 \end{pmatrix}$

(b) $\dfrac{d\mathbf{B}}{dt} = \begin{pmatrix} 6 & 0 \\ -1/t^2 & 4 \end{pmatrix}$

(c) $\displaystyle\int_0^1 \mathbf{A}(t)\,dt = \begin{pmatrix} \tan^{-1}t & \frac{3}{2}t^2 \\ \frac{1}{3}t^3 & \frac{1}{2}t^2 \end{pmatrix}\Bigg|_{t=0}^{t=1} = \begin{pmatrix} \frac{\pi}{4} & \frac{3}{2} \\ \frac{1}{3} & \frac{1}{2} \end{pmatrix}$

(d) $\displaystyle\int_1^2 \mathbf{B}(t)\,dt = \begin{pmatrix} 3t^2 & 2t \\ \ln t & 2t^2 \end{pmatrix}\Bigg|_{t=1}^{t=2} = \begin{pmatrix} 9 & 2 \\ \ln 2 & 6 \end{pmatrix}$

(e) $\mathbf{A}(t)\mathbf{B}(t) = \begin{pmatrix} 6t/(t^2+1)+3 & 2/(t^2+1)+12t^2 \\ 6t^3+1 & 2t^2+4t^2 \end{pmatrix}$

(f) $\dfrac{d}{dt}\mathbf{A}(t)\mathbf{B}(t) = \begin{pmatrix} (6-6t^2)/(t^2+1)^2 & -4t/(t^2+1)^2+24t \\ 18t^2 & 12t \end{pmatrix}$

(g) $\displaystyle\int_1^t \mathbf{A}(s)\mathbf{B}(s)\,ds = \begin{pmatrix} 6s/(s^2+1)+3 & 2/(s^2+1)+12s^2 \\ 6s^3+1 & 6s^2 \end{pmatrix}\Bigg|_{s=1}^{s=t}$

$$= \begin{pmatrix} 3t + 3\ln(t^2+1) - 3 - 3\ln 2 & 4t^3 + 2\tan^{-1}t - 4 - \pi/2 \\ (3/2)t^4 + t - (5/2) & 2t^3 - 2 \end{pmatrix}$$

33. $\begin{pmatrix} 1 & -1 & -5 & | & 7 \\ 5 & 4 & -16 & | & -10 \\ 0 & 1 & 1 & | & -5 \end{pmatrix} \implies \begin{pmatrix} 1 & -1 & -5 & | & 7 \\ 0 & 1 & 1 & | & -5 \\ 0 & 9 & 9 & | & -45 \end{pmatrix} \implies \begin{pmatrix} 1 & 0 & -4 & | & 2 \\ 0 & 1 & 1 & | & -5 \\ 0 & 0 & 0 & | & 0 \end{pmatrix}$

Letting $z = t$ we find $y = -5 - t$, and $x = 2 + 4t$.

36. $\begin{pmatrix} 1 & 0 & 2 & | & 8 \\ 1 & 2 & -2 & | & 4 \\ 2 & 5 & -6 & | & 6 \end{pmatrix} \implies \begin{pmatrix} 1 & 0 & 2 & | & 8 \\ 0 & 2 & -4 & | & -4 \\ 0 & 5 & -10 & | & -10 \end{pmatrix} \implies \begin{pmatrix} 1 & 0 & 2 & | & 8 \\ 0 & 1 & -2 & | & -2 \\ 0 & 0 & 0 & | & 0 \end{pmatrix}$

Letting $z = t$ we find $y = -2 + 2t$, and $x = 8 - 2t$.

39. $\begin{pmatrix} 1 & 2 & 4 & | & 2 \\ 2 & 4 & 3 & | & 1 \\ 1 & 2 & -1 & | & 7 \end{pmatrix} \implies \begin{pmatrix} 1 & 2 & 4 & | & 2 \\ 0 & 0 & -5 & | & -3 \\ 0 & 0 & -5 & | & 5 \end{pmatrix} \implies \begin{pmatrix} 1 & 2 & 0 & | & -2/5 \\ 0 & 0 & 1 & | & 3/5 \\ 0 & 0 & 0 & | & 8 \end{pmatrix}$

There is no solution.

42. $\begin{bmatrix} 2 & 4 & -2 & | & 1 & 0 & 0 \\ 4 & 2 & -2 & | & 0 & 1 & 0 \\ 8 & 10 & -6 & | & 0 & 0 & 1 \end{bmatrix} \xrightarrow[\text{operations}]{\text{row}} \begin{bmatrix} 1 & 2 & -1 & | & \frac{1}{2} & 0 & 0 \\ 0 & 1 & -\frac{1}{3} & | & \frac{1}{3} & -\frac{1}{6} & 0 \\ 0 & 0 & 0 & | & -2 & -1 & 1 \end{bmatrix}$; **A** is singular.

45. $\begin{bmatrix} 1 & 2 & 3 & 1 & | & 1 & 0 & 0 & 0 \\ -1 & 0 & 2 & 1 & | & 0 & 1 & 0 & 0 \\ 2 & 1 & -3 & 0 & | & 0 & 0 & 1 & 0 \\ 1 & 1 & 2 & 1 & | & 0 & 0 & 0 & 1 \end{bmatrix} \xrightarrow[\text{operations}]{\text{row}} \begin{bmatrix} 1 & 2 & 3 & 1 & | & 1 & 0 & 0 & 0 \\ 0 & 1 & \frac{5}{2} & 1 & | & \frac{1}{2} & \frac{1}{2} & 0 & 0 \\ 0 & 0 & 1 & -\frac{2}{3} & | & \frac{1}{3} & -1 & -\frac{2}{3} & 0 \\ 0 & 0 & 0 & 1 & | & -\frac{1}{2} & 1 & \frac{1}{2} & \frac{1}{2} \end{bmatrix}$

$\xrightarrow[\text{operations}]{\text{row}} \begin{bmatrix} 1 & 0 & 0 & 0 & | & -\frac{1}{2} & -\frac{2}{3} & -\frac{1}{6} & \frac{7}{6} \\ 0 & 1 & 0 & 0 & | & 1 & \frac{1}{3} & \frac{1}{3} & -\frac{4}{3} \\ 0 & 0 & 1 & 0 & | & 0 & -\frac{1}{3} & -\frac{1}{3} & \frac{1}{3} \\ 0 & 0 & 0 & 1 & | & -\frac{1}{2} & 1 & \frac{1}{2} & \frac{1}{2} \end{bmatrix}$; $\mathbf{A}^{-1} = \begin{bmatrix} -\frac{1}{2} & -\frac{2}{3} & -\frac{1}{6} & \frac{7}{6} \\ 1 & \frac{1}{3} & \frac{1}{3} & -\frac{4}{3} \\ 0 & -\frac{1}{3} & -\frac{1}{3} & \frac{1}{3} \\ -\frac{1}{2} & 1 & \frac{1}{2} & \frac{1}{2} \end{bmatrix}$

48. We solve

$$\det(\mathbf{A} - \lambda\mathbf{I}) = \begin{vmatrix} 2-\lambda & 1 \\ 2 & 1-\lambda \end{vmatrix} = \lambda(\lambda - 3) = 0.$$

For $\lambda_1 = 0$ we have

$$\begin{pmatrix} 2 & 1 & | & 0 \\ 2 & 1 & | & 0 \end{pmatrix} \Longrightarrow \begin{pmatrix} 1 & 1/2 & | & 0 \\ 0 & 0 & | & 0 \end{pmatrix}$$

so that $k_1 = -\frac{1}{2}k_2$. If $k_2 = 2$ then

$$\mathbf{K}_1 = \begin{pmatrix} -1 \\ 2 \end{pmatrix}.$$

For $\lambda_2 = 3$ we have

$$\begin{pmatrix} -1 & 1 & | & 0 \\ 2 & -2 & | & 0 \end{pmatrix} \Longrightarrow \begin{pmatrix} 1 & -1 & | & 0 \\ 0 & 0 & | & 0 \end{pmatrix}$$

so that $k_1 = k_2$. If $k_2 = 1$ then

$$\mathbf{K}_2 = \begin{pmatrix} 1 \\ 1 \end{pmatrix}.$$

51. We solve

$$\det(\mathbf{A} - \lambda\mathbf{I}) = \begin{vmatrix} 5-\lambda & -1 & 0 \\ 0 & -5-\lambda & 9 \\ 5 & -1 & -\lambda \end{vmatrix} = \begin{vmatrix} 4-\lambda & -1 & 0 \\ 4-\lambda & -5-\lambda & 9 \\ 4-\lambda & -1 & -\lambda \end{vmatrix} = \lambda(4-\lambda)(\lambda+4) = 0.$$

If $\lambda_1 = 0$ then

$$\begin{pmatrix} 5 & -1 & 0 & | & 0 \\ 0 & -5 & 9 & | & 0 \\ 5 & -1 & 0 & | & 0 \end{pmatrix} \implies \begin{pmatrix} 1 & 0 & -9/25 & | & 0 \\ 0 & 1 & -9/5 & | & 0 \\ 0 & 0 & 0 & | & 0 \end{pmatrix}$$

so that $k_1 = \frac{9}{25}k_3$ and $k_2 = \frac{9}{5}k_3$. If $k_3 = 25$ then

$$\mathbf{K}_1 = \begin{pmatrix} 9 \\ 45 \\ 25 \end{pmatrix}.$$

If $\lambda_2 = 4$ then

$$\begin{pmatrix} 1 & -1 & 0 & | & 0 \\ 0 & -9 & 9 & | & 0 \\ 5 & -1 & -4 & | & 0 \end{pmatrix} \implies \begin{pmatrix} 1 & 0 & -1 & | & 0 \\ 0 & 1 & -1 & | & 0 \\ 0 & 0 & 0 & | & 0 \end{pmatrix}$$

so that $k_1 = k_3$ and $k_2 = k_3$. If $k_3 = 1$ then

$$\mathbf{K}_2 = \begin{pmatrix} 1 \\ 1 \\ 1 \end{pmatrix}.$$

If $\lambda_3 = -4$ then

$$\begin{pmatrix} 9 & -1 & 0 & | & 0 \\ 0 & -1 & 9 & | & 0 \\ 5 & -1 & 4 & | & 0 \end{pmatrix} \implies \begin{pmatrix} 1 & 0 & -1 & | & 0 \\ 0 & 1 & -9 & | & 0 \\ 0 & 0 & 0 & | & 0 \end{pmatrix}$$

so that $k_1 = k_3$ and $k_2 = 9k_3$. If $k_3 = 1$ then

$$\mathbf{K}_3 = \begin{pmatrix} 1 \\ 9 \\ 1 \end{pmatrix}.$$

54. We solve

$$\det(\mathbf{A} - \lambda\mathbf{I}) = \begin{vmatrix} 1-\lambda & 6 & 0 \\ 0 & 2-\lambda & 1 \\ 0 & 1 & 2-\lambda \end{vmatrix} = \begin{vmatrix} 1-\lambda & 6 & 0 \\ 0 & 3-\lambda & 3-\lambda \\ 0 & 1 & 2-\lambda \end{vmatrix} = (3-\lambda)(1-\lambda)^2 = 0.$$

For $\lambda = 3$ we have

$$\begin{pmatrix} -2 & 6 & 0 & | & 0 \\ 0 & 0 & 0 & | & 0 \\ 0 & 1 & -1 & | & 0 \end{pmatrix} \implies \begin{pmatrix} 1 & 0 & -3 & | & 0 \\ 0 & 1 & -1 & | & 0 \\ 0 & 0 & 0 & | & 0 \end{pmatrix}$$

so that $k_1 = 3k_3$ and $k_2 = k_3$. If $k_3 = 1$ then

$$\mathbf{K}_1 = \begin{pmatrix} 3 \\ 1 \\ 1 \end{pmatrix}.$$

For $\lambda_2 = \lambda_3 = 1$ we have

$$\begin{pmatrix} 0 & 6 & 0 & | & 0 \\ 0 & 1 & 1 & | & 0 \\ 0 & 1 & 1 & | & 0 \end{pmatrix} \implies \begin{pmatrix} 0 & 1 & 0 & | & 0 \\ 0 & 0 & 1 & | & 0 \\ 0 & 0 & 0 & | & 0 \end{pmatrix}$$

so that $k_2 = 0$ and $k_3 = 0$. If $k_1 = 1$ then

$$\mathbf{K}_2 = \begin{pmatrix} 1 \\ 0 \\ 0 \end{pmatrix}.$$

57. Let

$$\mathbf{A} = \begin{pmatrix} a_{11} & a_{12} \\ a_{21} & a_{22} \end{pmatrix}.$$

Then

$$\frac{d}{dt}[\mathbf{A}(t)\mathbf{X}(t)] = \frac{d}{dt} \begin{pmatrix} a_1 & a_2 \\ a_3 & a_4 \end{pmatrix} \begin{pmatrix} x_1 \\ x_2 \end{pmatrix} = \frac{d}{dt} \begin{pmatrix} a_1 x_1 + a_2 x_2 \\ a_3 x_1 + a_4 x_2 \end{pmatrix} = \begin{pmatrix} a_1 x_1' + a_1' x_1 + a_2 x_2' + a_2' x_2 \\ a_3 x_1' + a_3' x_1 + a_4 x_2' + a_4' x_2 \end{pmatrix}$$

$$= \begin{pmatrix} a_1 & a_2 \\ a_3 & a_4 \end{pmatrix} \begin{pmatrix} x_1' \\ x_2' \end{pmatrix} + \begin{pmatrix} a_1' & a_2' \\ a_3' & a_4' \end{pmatrix} \begin{pmatrix} x_1 \\ x_2 \end{pmatrix} = \mathbf{A}(t)\mathbf{X}'(t) + \mathbf{A}'(t)\mathbf{X}(t).$$

60. Since

$$(\mathbf{AB})(\mathbf{B}^{-1}\mathbf{A}^{-1}) = \mathbf{A}(\mathbf{BB}^{-1})\mathbf{A}^{-1} = \mathbf{AIA}^{-1} = \mathbf{AA}^{-1} = \mathbf{I}$$

and

$$(\mathbf{B}^{-1}\mathbf{A}^{-1})(\mathbf{AB}) = \mathbf{B}^{-1}(\mathbf{A}^{-1}\mathbf{A})\mathbf{B} = \mathbf{B}^{-1}\mathbf{IB} = \mathbf{B}^{-1}\mathbf{B} = \mathbf{I}$$

we have

$$(\mathbf{AB})^{-1} = \mathbf{B}^{-1}\mathbf{A}^{-1}.$$